Cognitive Vehicular
Networks

Cognitive Vehicular Networks

Editors

Anna Maria Vegni
Department of Engineering (Section of Applied Electronics)
Roma TRE University, Rome, Italy

and

Dharma P. Agrawal
CDMC, EECS Department
University of Cincinnati, Cincinnati, OH, USA

CRC Press
Taylor & Francis Group
Boca Raton London New York

CRC Press is an imprint of the
Taylor & Francis Group, an **Informa** business

A SCIENCE PUBLISHERS BOOK

CRC Press
Taylor & Francis Group
6000 Broken Sound Parkway NW, Suite 300
Boca Raton, FL 33487-2742

© 2016 by Taylor & Francis Group, LLC
CRC Press is an imprint of Taylor & Francis Group, an Informa business

No claim to original U.S. Government works

Printed on acid-free paper
Version Date: 20151027

International Standard Book Number-13: 978-1-4987-2191-2 (Hardback)

Library of Congress Cataloging-in-Publication Data

Names: Vegni, Anna Maria, author. | Agrawal, Dharma P. (Dharma Prakash), 1945-
Title: Cognitive vehicular networks / authors, Anna Maria Vegni and Dharma P. Agrawal.
Description: Boca Raton : Taylor & Francis, 2016. | Includes bibliographical references and index.
Identifiers: LCCN 2015040689 | ISBN 9781498721912 (hardcover : alk. paper)
Subjects: LCSH: Vehicular ad hoc networks (Computer networks) | Cognitive radio networks.
Classification: LCC TE228.37 .V38 2016 | DDC 388.3/12--dc23
LC record available at http://lccn.loc.gov/2015040689

Visit the Taylor & Francis Web site at
http://www.taylorandfrancis.com

and the CRC Press Web site at
http://www.crcpress.com

Dedication

To my daughter Giulia—Anna Maria Vegni
To my grandson Dhevan—Dharma P. Agrawal

Preface

Overview and Goals

The use of Cognitive Radio (CR) has been advocated primarily for letting secondary users explore the possibility of utilizing unused channels by the primary users, thereby limiting the scope of its usefulness. But, recent developments made in the field of vehicular networks have enabled a new class of on-board entertainment and safety systems, by exploiting opportunistic spectrum usage by means of CR technology. As a result, the Cognitive Vehicular Networks (CVNs) provide additional spectrum opportunities outside the well-known 5.9 GHz band of IEEE 802.11p standard.

The CR technology is an enabling technology for opportunistic spectrum use, which directly benefits various forms of vehicular communication. A node in a CVN implements spectrum management functionalities, like detecting spectrum opportunities over frequency bands, selecting the channel to use based on the QoS requirements of the applications, and then transmitting data without causing any harmful interference to the licensed owners of the spectrum. CVNs have many unique characteristics such as the dynamicity over time of the spectrum availability, as a function of the activities of the licensed or primary users, as well as on the relative motion among them. For instance, at busy hours or in urban areas, spectrum information can be exchanged over multiple cooperating vehicles, leading to learn more about the spectrum availability.

This book focuses on CVNs as a new class of mobile ad hoc networks, which exploits cognitive aspects in the field of vehicular communications. The aim of the book is to provide a comprehensive guide to selected topics, both ongoing and emerging, in the area of CVNs, by using a treatment approach suitable for pedagogical purposes. This book is expected to address the topic of cognitive vehicular networks, with main features and key aspects. For instance, the use of resources in vehicular cognitive networks is investigated, as well as data dissemination techniques, while reducing packet retransmission (i.e., limitation of broadcast storm effect).

Selected topics that are covered in this book are related to: cognitive radio techniques, spectrum sensing approaches applied to vehicular networks, spectrum allocation techniques in CVNs, cognitive routing techniques in CVNs, data dissemination in CVNs, security in CVNs, as well as novel trends and challenges in CVNs.

Contributions come from worldwide well-known and high profile researchers in their respective specialities. The book is written to target both the student and the research community. We hope that the book will be a valuable reference for graduate level students and professional researchers, working in the ICT (Information and Communications Technology) domain and having extensive knowledge about many of the topics in the book, as well as people from industry, working on CR technologies applied to vehicular networks. Furthermore, it provides a comprehensive guide to selected topics, both ongoing and emerging, in the area of cognitive vehicular networks, by using a treatment approach suitable for pedagogical purposes. Finally, despite the fact that there are different authors for each chapter, we have made sure that the book materials are as coherent and synchronized as possible in order to be easy to follow by all readers.

Features of the Book

There are several reasons why this book can be a useful resource for both graduate students and researchers. We summarize below some of the important features of the book, such as:

- There are several specific journals that target for promotion or review of CR-based mobile networks. However, this book is the first about the application of CR in vehicular ad hoc networks. The topic of CR has attracted many researchers, and the usage of CR-based technique in vehicular networks is something of new that needs attention;
- Vehicular ad hoc networks is a research field that still needs contributions to address existing challenges, such as the use of resources through new technologies like CR-based solutions;
- We have attempted to write a clear and simple reference directed to any reader's level. Cognitive Vehicular Networks are a specific class of mobile networks that exploit the usage of CR-based approaches, then show challenges related to networking, as well as signal processing;
- This book offers a study of CVNs with the main features and fundamentals, starting from an introduction about CR technology in traditional networks up to the case of vehicular ad-hoc networks, and then investigates the main applications of CVNs.

Organization and Scanning of Chapters

The book is organized into 10 chapters, with each chapter written by topical area expert(s). The chapters are grouped into 3 parts.

Part I, namely *Cognitive Radio*, is devoted to topics on a brief introduction on cognitive networks, with particular attention to the case of vehicular networks. Chapter 1 surveys and reviews new approaches related to the synergistic combination of the CR with the opportunistic networks, specifically the vehicular networks. The background on features and peculiarities of the CR technology as a dynamic management of the spectrum access is provided. The CR technology is investigated in an opportunistic cognitive network environment, with a particular emphasis on data dissemination issues in vehicular networks.

Chapter 2 introduces the concept of vehicular ad hoc networks (VANETs), and investigates the existing communication methods, the cognitive radio technology, and its applicability to vehicular scenarios, along with a brief introduction to the advancement and research issues in the cognitive vehicular networks.

Part II, namely *Cognitive Radio for Vehicular Networks*, is devoted to introduce the concept of CVNs, and all the related aspects, such as spectrum allocation techniques, cognitive routing, data dissemination and connectivity analysis. This part is comprised of five chapters, i.e., Chapter 3, Chapter 4, Chapter 5, Chapter 6, and Chapter 7.

Chapter 3 investigates spectrum allocation techniques for vehicular CR networks. Indeed, a vehicular network may have variable contact rates, adjustable hop-to-hop and end-to-end delays, inconstant traffic density, and even adaptable goals, given by each service or application. In such complex heterogeneous and dynamic scenarios, spectrum allocation represents a challenge. For this aim, the authors present two policy-based solutions that allow cognitive vehicular networks to adapt distributively by merely instantiating few policies.

In Chapter 4 the authors present an overview about important technologies used to allow data dissemination in cognitive vehicular networks, i.e., data dissemination and communication approaches, broadcasting protocols, and main related applications.

Chapter 5 introduces the concept of CoVanets, CoRoute, and CoCast. CoVanets is a Cognitive VANET featuring cognitive radios specifically designed for the urban scenario. CoRoute is an anypath routing protocol that trades off robustness (to primary interference) with path length, and exploits multiple path diversity. Then, CoCast (Cognitive Multicast) is introduced for popular applications downloading, offering innovative features such as parallel data block transmissions in OFDM sub-channels and network coding across parallel channels.

Chapter 6 presents the topic of routing in cognitive vehicular networks. Initial research on routing for CVNs was mostly an extension to routing schemes proposed for VANETs, where the environment is extremely dynamic, due to high mobility of the vehicles and locality of spectrum availability. The main routing solutions for CVNs are then presented.

Chapter 7 presents an overview of network connectivity problem in CVNs. Due to mobility of the vehicles and the variable usage of channels by primary users, dramatic changes in spatial and temporal behaviors of the network topology occur. As a result, establishing continuous communication and disseminating online information result with intermittent connectivity, that causes degradations both in the vehicle satisfaction and communication quality.

Part III, namely *Applications for CVNs*, investigates the main solutions for applications in CVNs, starting from the topic of security and privacy up to data management strategies. Part III is comprised of three chapters, i.e., Chapter 8, Chapter 9, and Chapter 10.

Chapter 8 characterizes security and privacy issues in CVNs and discusses solutions in achieving secure cognitive communications. It reviews how security threats have been raised along with new architectures of vehicular networks.

Chapter 9 presents GroupConnect, a recent wireless technology showing the main features of mobile clouds but also pools resources of an entire group into a single virtual wireless node. GroupConnect is applied to the context of vehicular networks, by introducing vehicular clouds, defined as cloud services running on top of networks of vehicles. Finally, Chapter 10 concludes the book by means of an experimental evaluation of cognitive based solution for vehicular networks. A simulation approach is presented, and it is focused on the evaluation of data management strategies for vehicular networks by exploiting the use of mobile agent technology.

Target Audience

As stated early, this book is designed, in structure and content, with the intention to make it useful and readable at all learning levels. The book targets both the students and the research community. We face to graduate level students and professional researchers, working in the ICT domain and having extensive knowledge about many of the topics in the book, as well as people from the industry, working on cognitive radio technologies applied to vehicular networks.

The book will provide significant technical and practical insights in different aspects, starting from a basic background on cognitive radio, the inter-related technologies and application to vehicular networks, the technical challenges, implementation and future trends.

Acknowledgements

We are thankful to all the authors of the chapters of this book, who have worked very hard to bring forward this unique resource on Cognitive Vehicular Networks, intended to help students, instructors, researchers, and community practitioners.

Finally, we would like to sincerely thank our respective families, for their continuous support and encouragement during the course of this project.

Anna Maria Vegni
Dharma P. Agrawal

Contents

List of Figures

CHAPTER 1

CHAPTER 2

CHAPTER 3

CHAPTER 4

CHAPTER 5

CHAPTER 6

CHAPTER 7

CHAPTER 8

CHAPTER 9

CHAPTER 10

List of Tables

List of Contributors

Dharma P. Agrawal
University of Cincinnati, Cincinnati OH, USA.

Ozgur B. Akan
Faculty member, Director of Next-Generation Wireless Communication Laboratory, Koc University, Istanbul, Turkey.

Andre L.L. Aquino
Sensor Net-UFAL Research Group, Computer Institute, Federal University of Alagoas – Brazil.
Email: alla.lins@pq.cnpq.br; alla@ic.ufal.br

Elif Bozkaya
Department of Computer Engineering, Istanbul Technical University, Istanbul, Turkey.
Email: bozkayae@itu.edu.tr

Berk Canberk
Department of Computer Engineering, Istanbul Technical University, Istanbul, Turkey.
Email: canberk@itu.edu.tr

Donato Di Paola
Institute of Intelligent Systems for Automation, National Research Council (ISSIA-CNR), Bari, Italy.

Ozgur Ergul
Research & Teaching Assistant, Next-Generation Wireless Communication Laboratory, Koc University, Istanbul, Turkey.
Email: ozergul@ku.edu.tr

Flavio Esposito
Advanced Technology Group, Exegy, Inc., St. Louis, USA.
Email: flavio@bu.edu

Yi Gai
Intel Labs, Hillsboro, OR 97124, USA.
Email: yigaiee@gmail.com; yi.gai@intel.com

Mario Gerla
Computer Science, University of California, Los Angeles, CA, USA.
Email: gerla@cs.ucla.edu

Sergio Ilarri
University of Zaragoza, Spain.

Wooseong Kim
Computer Engineering, Gachon University, South Korea.
Email: wooseong.kim@gmail.com

Bhaskar Krishnamachari
Department of Electrical Engineering, University of Southern California, Los Angeles, CA 90089, USA.
Email: bkrishna@usc.edu

David H.S. Lima
Sensor Net-UFAL Research Group, Computer Institute, Federal University of Alagoas – Brazil.
Email: dhs.lima@gmail.com

Jian Lin
Department of Electrical and Computer Engineering, Georgia Institute of Technology, Atlanta, GA, 30332, USA.
Email: jlin61@gatech.edu

Valeria Loscrì
INRIA Lille-Nord Europe, Lille, France.
Email: valeria.loscri@inria.fr

Ahmad A. Mostafa
The British University in Egypt (BUE), Egypt.
Email: Ahmad.mostafa@bue.edu.eg

Heitor S. Ramos
Sensor Net-UFAL Research Group, Computer Institute, Federal University of Alagoas – Brazil.
Email: heitor@ic.ufal.br

Rhudney Simões
Sensor Net-UFAL Research Group, Computer Institute, Federal University of Alagoas – Brazil.
Email: rhudney.simoes@gmail.com

Oscar Urra
University of Zaragoza (Zaragoza, Spain).
Email: ourra@itainnova.es

Anna Maria Vegni
Roma TRE University, Rome, Italy.

Marat Zhanikeev
Department of Artificial Intelligence, Computer Science and Systems Engineering, Kyushu Institute of Technology, Fukuoka Prefecture, Japan.
Email: maratishe@gmail.com

Part I

Cognitive Radio

1

Opportunistic Cognitive Networks

Valeria Loscrí,[1,*] *Anna Maria Vegni*[2] and
Dharma P. Agrawal[3]

ABSTRACT

In the context of vehicular networks a very interesting application is represented by the rapid dissemination of messages, especially in the case of emergency and safety applications. Many issues and challenges are related to data dissemination in a dynamic environment, such as vehicular networks. Indeed, in such a high dynamic context, it is fundamental to guarantee fast reception of messages to all relevant vehicles, in order to make people aware of dangerous situations and allow them, for example, to change their routes to destinations in a timely fashion.

Very recently, the use of the Cognitive Radio (CR) paradigm is being exploited for building Vehicular Ad hoc NETworks (VANETs) to acquire more importance. The CR technology is foreseen as a very effective tool to improve the communication efficiency in vehicular networks. It is envisioned that future vehicles will be CR-enabled, and that the CR paradigm in the context of VANETs has a great potential to become a killer application due to the huge consumer market for vehicular communications. The need to be effective is also "justified" by considering the increasing number and varieties of Intelligent Transportation Systems (ITS) applications and different Quality of Service (QoS) associated with multiplicity of applications. As a result, the CR paradigm can be applied to improve the QoS parameters by simultaneously reducing wireless connectivity cost in the context of VANETs.

[1] INRIA Lille-Nord Europe, Lille, France.
[2] Roma TRE University, Rome, Italy.
[3] University of Cincinnati, Cincinnati OH, USA.
* Corresponding author: valeria.loscri@inria.fr

This chapter surveys and reviews new approaches related to the synergistic combination of the CR with the vehicular networks. We provide background on features and peculiarities of the CR technology as a dynamic management of the spectrum access and the adaptive Software Defined Radio (SDR). After describing the CR technology and how it is usable for VANETs, we present an overview of the main data dissemination mechanisms exploiting CR approaches in order to improve the data communication efficiency and reliability. Moreover, a comparison between traditional data dissemination solutions and techniques based on CR paradigm is also provided. Finally, we describe the issues and challenges related to the implementation of CR technology in vehicular networks, and then we provide the most recent advances and open research directions focusing on data dissemination in CR-VANETs.

1. Introduction to Cognitive Radio

The concept of CR [10, 14] can be summarized as a device equipped with a capability to observe and learn from the operating environment, and adapt their wireless communication parameters in order to optimize their network performance. The key attributes of cognitive radio are (i) to learn, (ii) to sense, and (iii) to adapt.

The learning mechanisms are necessary to acquire information about communication parameters and to capture any underutilized spectrum by sensing the surroundings. Adaptive and dynamic adjustment of the transmission parameters (i.e., transmission power) allows achieving better utilization of the spectrum. Future wireless communications will require an increasingly opportunistic use of the licensed radio frequency spectrum, and the CR paradigm provides an appropriate framework for this. The main objective of the CR scheme is to improve the spectrum usage efficiency, while minimizing the problem of spectrum overcrowding [22]. Indeed, a CR system is based on the fact that different parts of the channel are not only allocated to fixed and pre-assigned users, but idle segments of the spectrum are made available to other users.

Recent advances in Cognitive Radio Networks (CRNs) can deal with multi-hop networks, representing a promising design to leverage full potential of CRNs. One of the main features of multi-hop networks is the routing metric used to select the best route for forwarding the packets. In [27], Youssef et al. surveyed the state-of-the-art routing metrics for CRNs. The problem of reliability of a multi-hop and multichannel CRN has been investigated by Pal et al. in [15]. In such a scenario, to support multilink operations and networking functions, traditional spectrum sensing is not enough, and there is a need for developing a CR tomography to meet the general needs of networking operations. This topic has also been investigated by Kai-Yu in [20]. Well-designed multi-hop CRNs can provide enhanced bandwidth efficiency by using dynamic spectrum access technologies, as well as provide extended coverage and ubiquitous connectivity for end users. Prior to cognitive radio approaches, a better usage of the under used spectrum bandwidth has been obtained by focusing on the scheduler at the Medium Access Level (MAC), as covered in [9] and [8], that improves assignment of the time slots for each user. The advent of CR represents a new open research direction that presents specific challenges and underlying issues.

In [16], Sengupta and Subbalakshmi surveyed the unique challenges and open research issues in the design of multi-hop CRNs; as they focus on MAC and network layers of the multi-hop CR protocol stack. They investigate the issues related to efficient spectrum sharing, optimal relay node selection, interference mitigation, end-to-end delay, and many others. The concept of spectrum sharing represents an effective method to fix the spectrum scarcity problem. Indeed, spectrum-sharing solutions allow unlicensed users to coexist with licensed users, under the condition of protecting the latter from interference. In [20], Stotas and Nallanathan investigate the throughput maximization in spectrum sharing CRNs by introducing a novel receiver and frame structure, and then deriving an optimal power allocation strategy that maximizes the capacity of the proposed CR system. Another survey on recent advances in CR is presented by Wang et al. in [25], while Naeem et al. [12] discuss the issue and suitable solutions for efficient resource allocation in cooperative CRNs, in order to meet the challenges of future wireless networks. The authors in [12] also highlight the use of power control in cooperative CRNs.

As outlined in [24], features of opportunistic networks fit well in the context of vehicular ad hoc networks, since the network topology changes very quickly due to high vehicle speed while the connectivity could be easily disrupted easily. Vehicle-to-Vehicle communications may not be the most appropriate scheme for data dissemination, and so in this chapter we consider an integration and synergic action of the Cognitive Radio paradigm as that can be a valid solution for data delivery purpose.

2. Cognitive Aspects in Opportunistic Networks

The concept of Opportunistic Networks (ONs) is not new. There have been several projects and research works that have addressed this topic. One of the most prominent contribution in terms of importance is represented by SOCIALNETS [18], that covers online social networks and opportunistic mobile peer-to-peer wireless networks. Development of protocols that allow data to travel through human social structures has been proposed that exploits opportunistic networking technologies.

In the European project "Opportunistic Networks and Cognitive Management Systems for Efficient Application Provision in the Future Internet" [13], the main objective has been in designing, developing and validating the concept of applying opportunistic networks and respective cognitive management systems for efficient application/service/content provisioning in the Future Internet. The project aims at efficiently responding to the request of new applications/services built on the concepts of social networking, so that conceive smart devices and real ubiquitous networks can be characterized by connectivity everywhere. The project also considers applications including management of critical infrastructures, environment (ecosystem) protection, product manufacturing, digital services each having its own bandwidth and service provision requirements.

Opportunistic Networking (ON) is one of the key paradigms that supports direct communication among devices in a mobile scenario. Due to the high volatility and dynamicity of information exchanged among mobile nodes and such nodes take decisions based on partial or incomplete knowledge, the development of effective and efficient data dissemination schemes is still an open issue.

Delay Tolerant Network (DTN) routing mechanism in mobile scenarios provides capabilities for nodes to opportunistically communicate with each other according to the existing environment. Thus, DTN paradigm requires nodes to cooperate and share data information about specific packet forwarding approaches. As an instance, the transmission of a message to a destination node occurs under the concept that the message can be opportunistically routed through relay nodes, under the assumption that each node is willing to participate in the forwarding process.

ONs represent a very volatile and dynamic networking environment. Main applications for ONs are social networking, emergency management, pervasive and urban sensing as the basic concept is to involve sharing of content amongst interested users. Despite the fact that nodes have limited resources, existing solutions for content sharing require that the nodes maintain and exchange large amount of status information, even though this limits the system scalability.

In order to cope with this problem, the authors [23] present a solution based on cognitive heuristics. Cognitive heuristics are functional models of the mental processes, widely studied in the cognitive psychology field. They describe the behavior of the brain when decisions have to be taken quickly in spite of availability of incomplete information. Nodes maintain an aggregated information based on observations of the encountered nodes. The aggregate status and a probabilistic decision process is the basis on which nodes apply cognitive heuristics to decide how to disseminate content items when they meet each other. These two features allow the proposed solution to drastically limit the state kept by each node, and dynamically adapt according to both the dynamics of item being diffused and the dynamically changing node interests.

In [11] Mordacchini et al. present algorithms based on well-established models in cognitive sciences in order to disseminate both data items and semantic information associated with them. Semantic information represents both meta-data associated with data items (e.g., tags associated with them), and meta-data describing the interests of the users (e.g., topics for which they would like to receive data items). Their solution exploits dissemination of semantic data about the users, and their interests in guiding the dissemination of corresponding data items. Both dissemination processes are based on models coming from the cognitive sciences field named cognitive heuristics which describe how humans organize information in their memory and exchange it during interactions based on partial and incomplete information.

In Vehicular Ad hoc NETworks (VANETs), ON is well exploited through the concept of *bridging* among vehicles traveling on constrained paths at different speeds and forming clusters [17]. Partitioning could be possible between vehicles traveling in the same direction of the roadway.

As known, VANETs belong to the family of Mobile Ad hoc NETworks (MANETs) with a particular feature that mobile nodes are vehicles and are able to communicate with each other via opportunistic wireless links. Vehicles travel on constrained paths (i.e., roads and highways) and can exchange safety and entertainment messages among neighboring vehicles. According to available connectivity links different communication modes are allowed in VANETs. For example, a vehicle can transmit traffic information messages to its neighbors via Vehicle-to-Vehicle (V2V) mode, while it can receive data from a traffic light, i.e., a Road Side Unit (RSU) via Vehicle-to-Infrastructure (V2I) approach. Of course, hybrid approaches based on both V2V and V2I communication modes are exploited, specially in those scenarios with high dynamic network topology, with a very quick change of connectivity links.

There are several wireless access technologies that could be used for vehicular communications. The most common technologies installed in on-board devices are the IEEE 802.11 and Wireless Wide Area Network interface cards, like Long Term Evolution (LTE) and Worldwide Interoperability for Microwave Access (WiMax), as well as Global Navigation Satellite System (GNSS) receiver for vehicle positioning and tracking. In particular, IEEE 802.11 *p* standard is intended to operate with the IEEE 1609 protocol suite, which provides the Wireless Access in Vehicular Environments (WAVE) protocol stacks.

Finally, short-range communications are also possible with Personal Area Networks, through Bluetooth technology. The driver is expected to be connected to the on-board computer as well as with other neighboring drivers. The connectivity of the drivers, as well as that of passengers, is guaranteed by means of innovative applications, like those based on social networking (e.g., car and traffic information sharing).

From the above considerations, we can state that VANETs present typical features, and the main differences from MANETs are given by dedicated applications. VANETs arise from the need to enhance traffic and reduce the number of road victims in accidents. Safety and traffic management applications are typical for VANETs, requiring timely and reliable message delivery as compared to a typical MANETs. As a consequence, VANETs fit well into the class of opportunistic networks, since messages are forwarded following *store-carry-and-forward* approach that assume to be messages are stored in a vehicle and quickly forwarded over an available wireless link. Connectivity links are then opportunistically exploited to forward the messages within the network.

The main motivation for combining ONs with efficient mechanism is to manage the radio spectrum such as the CR paradigm derived from the demand for novel and different services and having improved and expanded use of wireless with an increased efficiency in resource provisioning. The idea of the synergistic combination of the ONs and Cognitive Management Systems is well represented in Figure 1.

The authors in [19] identify several additional benefits that could be derived from the combination of these two elements:

- *Increased utilization of resources*: The temporary feature of the resources allocated for the new Opportunistic Networks allows enhancement of the efficiency by assigning underutilized resources for use by ONs;
- *Lower transmission power*: A better management of the spectrum in the context of opportunistic networks can save resources that could be used for other users. Any reduction in the power levels is always considered in the context of *green* network applications;
- *Added-value for end-users*: In some specific cases, when the infrastructure indicates congestion in terms of access or has some limited coverage, service providers are able to support applications/services by exploiting the opportunistic networks paradigm with the cognitive management system.

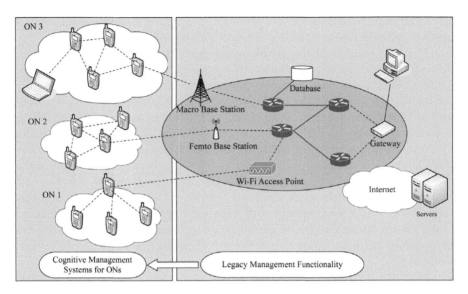

Figure 1. The synergistic combination of Opportunistic Networks and Cognitive Management Systems (CMSs) [19].

The cognitive aspect of the Opportunistic Network represents a very intriguing concept that normally has been considered mostly from a theoretical point of view with respect to architectures or routing approaches. A very interesting and novel perspective is taken in [19], where the authors consider opportunism as something that lies "in the fact that temporary opportunities of the radio environment—in terms of transmission and processing resources—are exploited to act as dynamic, ad hoc extensions of the infrastructure whenever and wherever needed."

Still from [19], the authors also consider the components of a Cognitive Management System as represented in Figure 2.

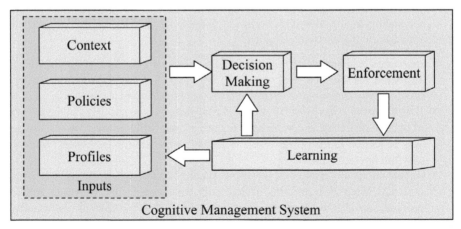

Figure 2. Components of a Cognitive Management System [19].

In Figure 3, a very simple example illustrates how the deployment of an Opportunistic Network can be effective. In fact, there can be users with poor channel quality towards the infrastructures (as shown in the left part of Figure 3) but with very good channel quality to the neighboring nodes. Compared to the users with better channel quality, the ones experiencing bad quality need more resources in terms of power and time to send the same amount of data. As reported in Figure 3, after the creation of the Opportunistic Network, the users with better channel quality can serve as in charge of the responsibility to forward the traffic to the poor users' channels. This will improve the overall capacity of the network by simultaneously decreasing the energy consumption.

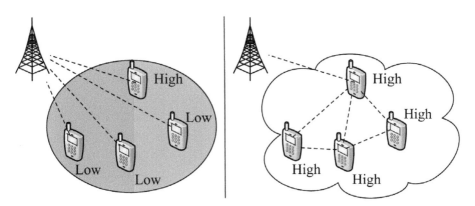

Figure 3. The impact of the use of an Opportunistic Network (ON): before (left) and after (right) [19].

3. Data Dissemination in Opportunistic Cognitive Networks

An important topic in opportunistic networks is represented by *data dissemination*. Indeed, this type of networks is characterized with very unstable topologies and there are several approaches for data forwarding. To address with this issue, many different directions can be considered. Several contributions propose different data-centric approaches for data dissemination where the data is proactively and cooperatively disseminated. In this type of networks, it can occur very frequently that sources and receivers are not aware of each other, and for this reason main data dissemination approaches are based on public/subscribe model. Finally, other kinds of proposals belong to the opportunistic data-dissemination techniques, based on social networks, gossiping, epidemic algorithms, etc. The common point of all these mechanisms is that there is no solution guaranteeing data delivery.

In Figure 4, a taxonomy for data dissemination approaches is shown. Ciobanu et al. remark the importance of the organization of the network as far as the data dissemination mechanism is concerned. They argue that different types of nodes can be categorized with different levels of centrality metrics. The presence of specific kinds of nodes such as *hubs*, belongs to the category of data dissemination algorithms with infrastructure. On the other hand, in the context of opportunistic networks, maintenance

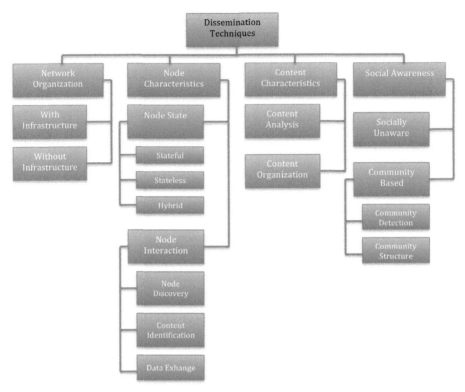

Figure 4. A taxonomy for data dissemination mechanisms [2].

of an infrastructure can be difficult and costly, since there are too many messages to be exchanged. Based on these considerations, most of the data dissemination mechanisms for opportunistic networks rely on a network without infrastructure. The authors outline the importance of the network organization as a key element that impacts on data transfer policies. Moreover, existing features of the nodes (i.e., node characteristics in Figure 4) directly influence data dissemination technique.

A comprehensive approach requires control traffic and in context of such an opportunistic network, where the topology changes occur very frequently, this mechanism can be very expensive. On the other hand, a stateless approach could result in very poor results in the case of event flooding. From the above considerations, it is natural to think about hybrid approaches that make possible the convergence of the advantages coming from these two approaches.

Conti et al. [3] propose some data dissemination heuristics in the context of opportunistic networks based on the exploitation of environmental information. The heuristics they propose are recognition based and can be used to perform an effective dissemination of the data. In the context of vehicular networks, by increasing the reliability of the data delivery the authors in [1] present a data dissemination mechanism that involves the closest vehicles, and thereby reduces the delay. Their approach is a predictive method based on a Hidden Markov Model, where availability of the channels for the subsequent time slots can be "predicted". This model allows a faster and more reliable allocation of channels.

A mechanism based on the exploitation of soft-input/soft-output data fusion in VANETs in combination with the Cognitive Radio paradigm has been proposed in [4]. The main objective of Cordeschi et al. is to perform traffic offloading towards Clouds, and to design and implement a distributed and adaptive resource management controller. This is realized by opportunistically acceding to a spectral-limited wireless backbone built up by multiple RSUs. In this way, on-board smartphones, that are typically limited both in terms of energy and computing capabilities, can access and utilize the V2I Wi-Fi connections and perform the traffic offloading to the Cloud. In order to accomplish this goal, Cordeschi et al. [4] model the resource management issue as a suitable constrained stochastic Network Utility Maximization problem.

An optimal cognitive resource management controller dynamically allocates the access time-windows at the serving RSUs, together with the access rates and traffic flows at the served Vehicular Clients (i.e., the secondary users of the wireless backbone). Notice that the controller provides hard reliability guarantees to the Cloud Service Provider (i.e., the primary user of the wireless backbone) on a per-slot basis, and is also capable of self-acquiring context information about the currently available bandwidth-energy resources. Through the use of the controller, the system quickly adapts to the mobility-induced abrupt changes of the state of the vehicular network, even in the presence of fadings, imperfect context information and intermittent V2I connectivity.

In [6], Kirsch et al. investigate on safety and travel efficiency issue in VANETs that are present when sharing information between vehicles and RSUs, specifically for scenarios with large amount of spectral congestion. Spectral congestion is possible when there is high vehicle density such as traffic jams. In [6], the authors propose a CR system to spatially and temporally add additional channels to VANETs. As a result, the additional spectrum can increase the throughput and provides a decrease of the probability of packet collisions. The use of spectrum sensing based solutions for an opportunistic spectrum access among neighboring vehicles is investigated in [7]. For the collaborative spectrum sensing, Belief Propagation (BP) is applied to tackle the distributed observations and to exploit redundancies in both space and time. Finally, in [5], Kakkasageri and Manvi investigate data aggregation in VANETs by means of a cognitive agent that exploits regression mechanisms. Specifically, the regression based cognitive agent approach efficiently aggregates the collected critical information and minimizes redundant data dissemination.

Furthermore, Sun et al. [21] present a Quality-Driven Adaptive Video Streaming for CR-based VANETs. It is obvious that channel conditions are highly dynamic due to both vehicle mobility and primary user activity. In order to support high-quality video playback in such a challenging scenario, the authors in [21] present an adaptive video streaming algorithm built on Scalable Video Coding (SVC); the proposed technique reduces interruption ratio and improves visual quality. The streaming algorithm is also capable of deciding an appropriate number of video layers for vehicle users, by taking into account several important factors like vehicle position, velocity, and the activity of primary users.

Finally, in [26], the authors address Popular Content Distribution (PCD) mechanisms in VANETs for the case of file downloaded by a group of on-board units (OBUs) passing by a RSU. Due to high speeds and channel fadings, the OBUs may not finish downloading the entire file from the RSU, but only possess several content pieces. As a solution, the authors in [26] propose a cooperative approach using coalitional graph game to establish a peer-to-peer (P2P) network among the OBUs, by means of a CR-based technique of V2V transmissions.

4. Conclusion

In this chapter, we presented the main aspects of an Opportunistic Network. This represents a very challenged type of network for several reasons, such as (i) the quality of the link is very variable, (ii) the contacts are extremely intermittent, (iii) the routing can be very challenging, since for most of the time, a path from source to destination may not exist, etc.

We considered these type of networks and we also treated the Vehicular Networks as a "special" type of ONs. In the context of ONs, it is necessary to improve wireless networking services provision that could play a primary role for data dissemination effectiveness. On the other hand, the CR paradigm has been conceived to address challenges related with a better management of radio spectrum resources, and also to respond to an increased need for wireless networking provisions. The combination of the ONs and the CR paradigm could play a very primary role in the effective management of the spectrum resources, in order to obtain an improved link stability. As a result, a better management will impact on one of the most important applications in the context of ONs and Vehicular Networks, namely the data dissemination.

In this chapter, we have analyzed different approaches in terms of data dissemination. This kind of application is very prominent and the research community is very active as shown by the several contributions in the vehicular context, where efficient data aggregation mechanisms have been investigated and proposed. This is also not surprising, given the critical data that can circulate above all in the VANETs.

References

[1]   I.H. Brahmi, S. Djahel and Y. Ghamri-Doudane. A hidden dissemination of safety messages in vanets. In IEEE Globecom, 2012.

[2]   R. Ciobanu and C. Dobre. Data dissemination in opportunistic networks. In Proc. of 18th International Conference on Control Systems and Computer Science (CSCS-18), 2011.

[3]   M. Conti, M. Mordacchini and A. Passarella. Data dissemination in opportunistic networks using cognitive heuristics. In IEEE International Symposium on a World of Wireless, Mobile and Multimedia Networks (WoWMom), 2011.

[4]   N. Cordeschi, D. Amendola, M. Shojafar and E. Baccarelli. Distributed and adaptive resource management in Cloud-assisted Cognitive Radio Vehicular Networks with hard reliability guarantees. Vehicular Communications, pp. 1–12, 2015.

[5]   M.S. Kakkasageri and S.S. Manvi. Regression-based critical information aggregation and dissemination in VANETs: A cognitive agent approach. Vehicular Communications, pp. 168–180, 2014.

[6]   N.J. Kirsch and B.M. O'Connor. Improving the performance of vehicular networks in high traffic density conditions with cognitive radios. In IEEE Intelligent Vehicles Symposium (IV), 2011.

[7]   H. Li and D.K. Irick. Collaborative spectrum sensing in cognitive radio vehicular ad hoc networks: Belief propagation on highway. In IEEE 71th Vehicular Technology Conference (VTC) Spring, 2010.

[8]   V. Loscri. A new distributed scheduling scheme for wireless mesh networks. In Proceedings of the IEEE 18th International Symposium on Personal, Indoor and Mobile Radio Communication (PIMRC), 2007.

[9]   V. Loscri. A queue dynamic approach for the coordinated distributed scheduler of the ieee 802.16. In Proceedings of the IEEE International Symposium on Computers and Communications (ISCC), pp. 1–6, 2008.

[10] J. Mitola and G.Q. Maguire, Jr. Cognitive radio: making software radios more personal. Personal Communications, pp. 13–18, 1999.

[11] M. Mordacchini, L. Valerio, M. Conti and A. Passarella. A cognitive-based solution for semantic knowledge and content dissemination in opportunistic networks. In IEEE International Symposium on a World of Wireless, Mobile and Multimedia Networks (WoWMom), 2013.

[12] M. Naeem, A. Anpalagan, M. Jaseemuddin and D.C. Lee. Resource allocation techniques in cooperative cognitive radio networks. Communications Survey Tutorials, pp. 729–744, 2014.

[13] OneFit. Project Website: http://www.ict-onefit.eu.

[14] P. Pace and V. Loscri. A step forward in the cognitive direction. In Proceedings of the 21st International Conference on Computers, Communications and Networks (ICCCN). IEEE, 2012.

[15] R. Pal, D. Idris, K. Pasari and N. Prasad. Characterizing reliability in cognitive radio networks. In Proceedings of the 1st International Symposium on Applied Sciences on Biomedical and Communication Technologies (ISABEL), pp. 1–6, 2008.

[16] S. Sengupta and K.P. Subbalakshmi. Open research issues in multi-hop cognitive radio networks. Communications Magazine, pp. 168–176, 2013.

[17] Y. Shao, C. Liu and J. Wu. Delay tolerant networks in VANETs, chapter Handbook on Vehicular Networks. M. Weigle and S. Olariu, Taylor & Francis.

[18] SOCIALNETS. Project Website: http://www.social-nets.eu.

[19] V. Stavroulaki, K. Tsagkaris, M. Logothetis, A. Georgakopolos, P. Demestichas, J. Gebert and M. Filo. An approach for exploiting cognitive radio networking technologies in the future internet. Vehicular Technology Magazine, 2011.

[20] S. Stotas and A. Nallanathan. Enhancing the capacity of spectrum sharing cognitive radio networks. Transactions on Vehicular Technology, pp. 3768–3779, 2014.

[21] L. Sun, A. Hunag, M. Shan, H. amd Xing and L. Cai. Quality-driven adaptive video streaming for cognitive vanets. In IEEE 80th Vehicular Technology Conference (VTC) Fall, 2014.

[22] Y. Tawk, J. Constantine and G.C. Christodoulou. Cognitive-radio and antenna functionalities: a tutorial. Antenna Propagation Magazine, pp. 231–243, 2014.

[23] L. Valerio, M. Conti, E. Pagani and A. Passarella. Autonomic cognitive-based data dissemination in opportunistic networks. In IEEE International Symposium on a World of Wireless, Mobile and Multimedia Networks (WoWMom), 2013.

[24] A.M. Vegni, C. Campolo, A. Molinaro and T.D.C. Little. Modeling of Intermittent Connectivity in Opportunistic Networks: The Case of Vehicular Ad hoc Networks. Routing in Opportunistic Networks, Springer Science+Business Media New York, USA, 2013.

[25] B. Wang and K.J.R. Liu. Advances in cognitive radio networks: a survey. Journal on Selected Topics Signal Processing, pp. 5–23, 2011.

[26] T. Wang, L. Song and Z. Hou. Collaborative data dissemination in cognitive vanets with sensing-throughput tradeoff. In IEEE 1st International Conference on Communications in China (ICCC), 2012.

[27] M. Youssef, M. Ibrahim, M. Abdelatif, C. Lin and A.V. Vasilakos. Routing metrics of cognitive radio networks: a survey. Communications Survey Tutorials, pp. 231–243, 2014.

2

Towards Cognitive Vehicular Networks

Ahmad A. Mostafa,[1,*] *Dharma P. Agrawal*[2] *and*
Anna Maria Vegni[3]

ABSTRACT

Vehicular networks are an essential part of today's research and tomorrow's reality. The ability to connect speedy vehicles does not only open endless possibilities for application advancements and enhancements to the vehicular experience, but it also helps avoid many dangers and issues that arise such as accidents, time and money wasted because of traffic, and many others. Previously, vehicular connectivity primarily relied on either vehicles communicating with each other, which are called Vehicle-to-Vehicle (V2V) communications, or vehicles communicating through network infrastructure on the roadside called Vehicle-to-Infrastructure (V2I) mode. The V2V technology relies on Dedicated Short-Range Communication (DSRC) channels, while V2I relies on the existing allocated bandwidth for the infrastructure technology.

Infrastructures utilized in the V2I communications are networks available on roads such as cellular networks, WiMAX or other telecommunication networks. However, due to bandwidth limitations of infrastructure networks and time delay performance requirements for vehicular communications, conventional infrastructure communications are no longer adequate for vehicular communication. Considering this current necessity, the field of cognitive vehicular networks has emerged. In this chapter, we introduce the concept of vehicular networks, the existing communication methods, the cognitive radio technology, and its applicability to vehicular networks, along with a brief introduction to the advancement and research issues in the cognitive vehicular networks.

---

[1] The British University in Egypt (BUE), Egypt.
[2] University of Cincinnati, Cincinnati OH, USA.
[3] Roma TRE University, Rome, Italy.
* Corresponding author: Ahmad.mostafa@bue.edu.eg

VANETs: Connecting Vehicles

Over the past few decades, the world has been rapidly embracing and enjoying the power of connectivity. Initially, this was due to the possibility and the promise of connecting detached machines through the Advanced Research Projects Agency Network (ARPANET) in the late 1960s, and then evolving into what we now know as the Internet. Although the focus of the Internet was providing connectivity to different users using unrelated end-machines, the past decade has witnessed further steps in this evolution.

During this decade, the focus of research has been geared towards connecting devices and things feasible through state-of-the-art networks. Some of these networks are based on infrastructure, such as many Internet of Things (IoT) products. However, infrastructure is not always a feasible option, especially in mobile and rapidly changing networks. In the latter case, *ad-hoc networks* have become the basis for such connectivity. Practical examples for such networks are those that connect vehicles, i.e., Vehicular Ad hoc Networks (VANETs).

With the advancements in the technology, size, and capabilities of sensors, memories, and processors, the road has been paved for the emergence of vehicular networking technology. Different types of sensors are now placed in many locations within the vehicle and on the road, and with communication ability; the data and information gathered from these sensors are being transferred from one vehicle to another. This capability for vehicles to both sense and communicate has empowered them with a new realm of applications and services that enhance the passenger vehicle experience. An example would be for the vehicles to sense their close proximity to detect for possible accidents and hence, they communicate with each other to avoid this occurrence. While the previous example was safety oriented, many other applications span from basic entertainment and convenience applications, to more important and crucial applications related to health, vehicle and road performance. These advancements and potential are shaping vehicular networks to be one of the most innovative and crucial paradigms in the recent years, and to become a backbone for many applications that are of boundless significance.

Applications based on VANETs are mainly categorized in one of the three categories, i.e., (i) safety applications, (ii) traffic and road conditions applications, or (iii) comfort and entertainment applications.

Safety applications focus on the safety of both the vehicles and the occupants, and hence cover various issues such as collision avoidance, vehicle components failure alerts, personal health such as vital measurements, and other safety oriented information. On the other hand, *traffic and road conditions applications* are aimed at improving the traveling experience for the users such as minimizing the wait time at a traffic light, or the traffic jam due to congestion. Moreover, the existence of road conditions such as bumps, potholes, incomplete constructions, and other obstacles can be effectively communicated among vehicles. Intelligent sensor systems installed into vehicles will eventually use this information, forwarded by other communicating vehicles in order to enhance the road trip experience and decide the best routes from a source to a destination. Finally, *comfort applications* are those that aim to provide entertainment and comforting applications for the users such as access to Internet, video streaming, collaborative video gaming, etc. Figure 1 depicts the different services that can be provided through vehicular technology [1].

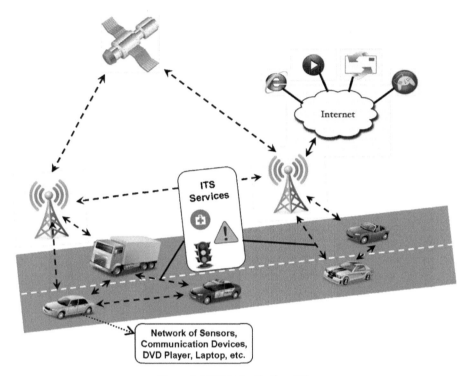

Figure 1. VANET Applications [1].

Each category of application has its own set of requirements that vary according to its importance. For example, health related applications require minimum time delay and latency. On the other hand, comfort and entertainment applications can tolerate some latency and can be modeled as Delay Tolerant Networks (DTN). Quality of Service (QoS) required by each application category has resulted in different technical requirements for the networks, which led to diverse methods of communication in vehicular networks. These methods are categorized into either Vehicle-to-Vehicle (V2V) or Vehicle-to-Infrastructure (V2I) communication.

V2V communications are primarily ad hoc networks that are formed among moving vehicles. The transmission range for V2V varies between several meters to hundreds of meters, due to different power levels at the transmitter, as well as environment conditions (i.e., urban or rural scenario). Due to the high mobility nature of the network, V2V communication performance faces many challenges in routing, which leads to a deteriorated end-to-end packet delivery delay. On the other hand, V2I communication establishes a connection between the vehicle and network infrastructure laying on the side of the road, whether it is cellular networks, or other kinds of infrastructure-based networks.

In this chapter we start from a description of the main features of VANETs, and then arrive at the concept of Cognitive Vehicular Networks. Specifically, leveraging on the main drawbacks of vehicular networks, the need to extend bandwidth requirements, and improve time delay performance requirements, we then introduce the cognitive vehicular networks. This chapter is organized as follows. Section 1 describes the vehicular networking technologies, the existing communication methods and main protocols. In Section 2 we introduce the concept of cognitive radio, while in Section 3 we describe the merge of cognitive radio technologies with the vehicular networking paradigm, along with a brief introduction to the advancement and research issues in cognitive vehicular networks. Finally, conclusions and discussion aspects are drawn at the end of this chapter.

1. Vehicular Networking Technologies

New vehicle technologies are composed of two main elements: (i) sensing, and (ii) wireless communication. The sensing part is responsible for understanding the environment, surroundings, state, and the occupants of the vehicle. The wireless communication element is the vehicular networking, which is responsible for communicating the information retrieved from the sensors wirelessly to the vehicle or entity that will be able to leverage and use that information.

Creating a network between different mobile devices without the use of infrastructure has been widely studied in research. These types of networks are called the Mobile Ad-Hoc Networks (MANETs). According to [2], MANETs are a network that is established when nodes come together only as needed. However, such a network is not necessarily formed with any support from any existing infrastructure or fixed stations. In these networks, nodes communicate in a multi-hop fashion, where network packets arrive to the destination from the source by hopping through other nodes in their vicinity. It has been proposed that vehicles use the same communication paradigm to communicate with each other, namely V2V communications.

However, there are fundamental differences between VANETs and a generic class of MANETs [3]. The first difference between both is the very high mobility of the nodes in VANETs. This causes a rapid and frequent change in the topology of the network, which leads to continuously changing routes from a source to destination vehicles. The rapid topology changes also result in a very short window of time for the communication to take place among vehicles, leading to a frequently disconnected network.

When it comes to routing, VANET nodes are defined by the geographical location, while in MANETs, nodes are defined by their IDs; hence routing in VANET is primarily geography-oriented. Moreover, in VANETs, routes and mobility are constrained and predictable according the roads, highways, traffic lights, stop signs, etc.

Finally, the propagation model in VANET networks is constrained between highway, city, or a rural area. In the city and rural environment, the signal reflection off either buildings; hills; dense forests or any other kind of opaque object is common, which is not the case in the highway environment. However, the vehicle density in the city model is significantly higher, which encourages taking advantage of multi-hop communication that also makes the system more prone to interference between the signals from the vehicles.

When it comes to large data processing power, memory, and even battery power, VANETs are adequately equipped in contrast to traditional MANETs. These differences hindered the adoption of the different MANET routing algorithms and communication methods to VANETs, and new paradigms and algorithms had to be introduced, or the old ones had to be appropriately modified.

The vehicle communication in V2V mode is based on the Dedicated Short-Range Communication (DSRC) channels, which is the name for an older technology, which has been used for vehicle to road communication such as tollgates. The channels used are in the 5.9 GHz band. The DSRC working group is working on the Wireless Access in Vehicular Environment (WAVE) standards. Currently, the US FCC has allocated a band of 75 MHz for Intelligent Transportation Services (ITS), which is an umbrella for the vehicular networks. Figure 2 explains the difference between the WAVE and the DSRC IEEE 802.11p standards, and what layers each is associated with.

The V2V technology is able to provide the quality of service required for few applications. However, this is not sufficient or adequate for some other applications, especially those that require high reliability with minimum delay. For these applications the use of infrastructure for communication would be required. Hence, the other type of communication in the vehicular networks is the V2I communication.

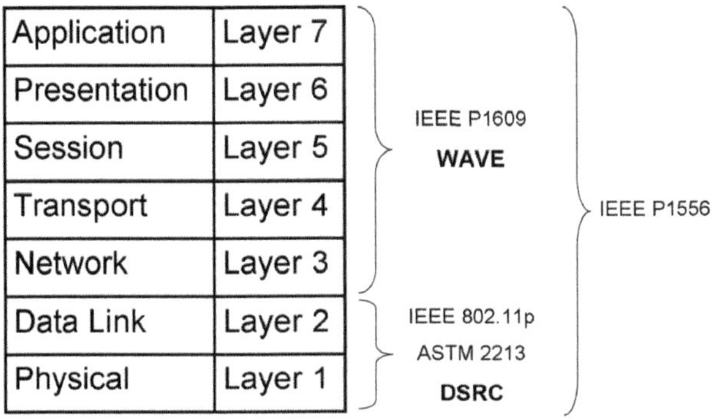

Application	Layer 7
Presentation	Layer 6
Session	Layer 5
Transport	Layer 4
Network	Layer 3
Data Link	Layer 2
Physical	Layer 1

IEEE P1609
WAVE

IEEE P1556

IEEE 802.11p
ASTM 2213
DSRC

Figure 2. Breakdown of standards and network layers for the vehicular communication technology [4].

In V2I, vehicles connect with each other or with the Internet through infrastructure located on the sides of the roads, namely Road Side Units (RSUs). The network infrastructure is centralized and can be easily managed. However, they come at a substantial financial cost, and can be easily overloaded. For example, the cellular network is already overloaded with data due to the wide adoption of smartphones, which resulted in dropped calls, spotty services, and delayed text [5]. This results in a poor quality of services, and hence, limits the ability to use a new technology on existing overloaded infrastructure networks.

Vehicular networks are assumed to combine both V2V as well as V2I technologies in a hybrid fashion, where they interchange between them based on the requirements and availability. For example, if a vehicle is traveling alone on a highway, it will resort to always communicate through the infrastructure on the side road. Figure 3 shows the difference between the three communication paradigms in vehicular networks [2].

While it is a common expectation for the vehicular network to rely completely on the freely available V2V wireless communication [13], the use of infrastructure has become a necessity due to underlying constraints. A rapid topology change leads to a disconnected network in V2V, while the vehicles remain in the communication range of the infrastructure for some time in V2I mode. This allows the vehicles to have a channel that is available for a short time to transmit and receive packets without any disruptions.

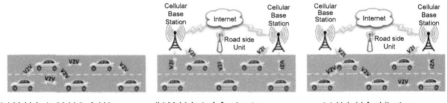

(a) Vehicle-to-Vehicle Ad Hoc Network

(b) Vehicle-to-Infrastructure Network

(c) Hybrid Architecture

Figure 3. Communication paradigms in vehicular networks [3].

Some solutions have been introduced to overcome the disadvantage of the overloaded status of current infrastructure networks. The Intelligence Transportation System (ITS) of the Department of Transportation (DoT) has proposed one of these solutions. They recommended installing completely new infrastructure to be used solely for the vehicular networks [14]. This proposed solution would cost approximately 251 million dollars to partially cover the roads in the United States. This is not a very feasible or practical solution, since many roads will not be covered by these infrastructures.

Recently, some other solutions have been proposed to overcome the issue of infrastructure overload without the need for the excessive cost. One of the most promising solutions proposed is the use of *cognitive radio* to access and use infrastructure in vehicular networks. In the next sections, we present cognitive radio technology, its use in vehicular networks, and some of the open research issues in the cognitive vehicular networks.

As mentioned earlier, the environmental characteristics and constraints introduced in vehicular networks are not similar to what has been introduced before in conventional MANETs. This lead to variations in the different protocols used and adapted for VANETs. In the following sections, we will present a summary of the protocols for the different networking layers in VANETs.

Physical Layer

The protocols for the physical layer in VANET rely on the DSRC technology. It operates in the range of 5.9 GHz band and supports communication with a data rate of up to 6 Mbps between vehicles moving at a speed up to 200 Km/h. The DSRC frequency band is usually divided into 7 channels, 6 used for service and one for control. The control channel is reserved for safety related and control messages, while the service channels are used for other types of data. DSRC is known as IEEE 802.11*p* WAVE.

MAC Layer

MAC protocols for VANET have to adapt as per several requirements for the vehicular environment. Some of these necessities are related to the quality of service required by the applications, such as the communication reliability and efficiency. However, some other requests are more general such as solving the hidden terminal problem. The DSRC technology is a member of the IEEE 802.11 family, hence, the MAC protocols for VANETs stem from those generally used in IEEE 802.11. For example, the WAVE technology uses the Carrier Sense Multiple Access with Collision Avoidance (CSMA/CA) as the medium access technique. Moreover, some other MAC protocols were introduced for VANETs, mainly the ADHOC MAC [6], and the D-MAC [7]. The ADHOC MAC protocol is based on the reliable reservation ALOHA protocol, while the D-MAC is based on the use of directional antenna that allows wireless communication and transmission to take place in a specific direction.

Network Layer

Due to the rapid changes in the network topology of VANETs caused by the vehicles' mobility, routing of packets remains to be one of the main challenges. Communication in VANETs is one of three categories, i.e., (i) unicast, (ii) multicast, and (iii) broadcast. The type of communication used depends mainly on the application requirements. For example, some applications require data or packets to be sent to a specific destination vehicle, and hence it will use unicast. Other applications will require data to be sent to any vehicle within a certain geographical area and thus, it will use multicast communication. If data is required to be sent to all vehicles in the network, broadcast communication will be used.

A significant number of routing algorithms have been developed for vehicular networks, and as with multi-hop wireless communication. These are either topology-based or position-based algorithms. Moreover, network layer protocols vary in their communication paradigm; the way they forward packets, the targeted architecture is V2V, V2I, or hybrid approach. Some of these protocols are tailored to specific application requirements and are appropriate for certain road scenarios such as city, highway, grid, or any other scenario. Table 1 shows a comparison between a few routing algorithms introduced in VANETs, and their associated drawbacks [2].

Transport Layer

The application delay requirements in VANET are very stringent, and the performance of the network rapidly deteriorates due to very quick topological changes, which only gives the vehicle a chance to communicate for a few seconds. This creates a hurdle for many applications that require communication to be reliable and in-order delivery.

Table 1. Routing protocols for VANET [3].

Routing Protocol	Communication Paradigm	Forwarding Strategy	Architecture	Scenario	Application
GPSR	Unicast	Greedy Forwarding	V2V	Real City Traces	CBR Traffic
GPCR	Unicast	Packet Forwarding	V2V	Real City Traces	-
VADD	Unicast	Packet Forwarding with Prediction	V2V	Real City Traces	CBR Traffic
A-STAR	Unicast	Packet Forwarding with Traffic Info	V2V	Real City Grid	CBR Traffic
CAR	Unicast	Packet Forwarding	V2V	Real City Traces	CBR Traffic
PROMPT	Unicast	Packet Forwarding with Position Based	V2I	Real City Traces	Variable Traffic
GyTAR	Unicast	Packet Forwarding with street awareness	V2V	Real City Traces	CBR Traffic
IVG	Geocast	Packet Forwarding	V2V	Urban and Rural Road Traces	-
Cashing Geocast	Geocast	Packet Forwarding with Caching	V2V	Random Traces	-
BROADCOMM	Broadcast	Packet Forwarding with virtual cells	V2V	Fixed Traces	Simple Broadcast

In regular wired networks, this is usually achieved by using TCP as the transport layer protocol. However, TCP performs poorly in wireless networks characterized by high node mobility and very frequent topology changes [8]. Some transport layer protocols have been proposed to overcome these drawbacks [9, 10, 11].

One of these protocols is the Ad-hoc TCP (ATCP) protocol [9] where the issues of packet loss, route changes, network partitions, multipath routing and packet reorder have been overcome [10]. The authors proposed a cross-layer mechanism that allows the network layer to receive feedback regarding the route status, and hence, adjusts the sender node status to either be in retransmission mode or in congestion control mode. Each mode has its specifications about the permissibility to transmit packets and the size of the congestion window.

In the Vehicular Transport Protocol (VTP), a protocol is proposed based on unicast routing [11]. The sender calculates statistical data based on ACK received or the lack of them, and gathers information from intermediate nodes, and concludes whether the VTP remains connected or in disrupted mode. VTP heavily relies on position based routing schemes.

Application Layer

Different application layer protocols have been proposed in the literature that is tailored specifically for VANETs. An example of these protocols is the Vehicular Information Transfer Protocol [12]. Due to the nature of the applications using the VANET, the protocols proposed have a focus on the minimization of end-to-end communication delay.

2. Cognitive Radio Technology

The rapid growth of wireless devices and applications has led to a parallel exponential growth in the wireless radio spectrum. For example, it is predicted that the mobile data usage will exceed what the existing networks infrastructure can handle [15]. In these networks, the bandwidth was utilized by statistically assigning the primary users a portion of the spectrum. However, upon further measurements and investigation, it was discovered that the primary users extremely underutilize the spectrum assigned to them. In one measurement, it was shown that for the spectrum below 6 GHz, over 60% of it is not being used or is underutilized [16].

The Working Group Frequency Management (WGFM) runs a few campaigns to monitor the usage of the frequency spectrum. In 2009, it produced the results that can be seen in Figure 4, showing the utilization for the frequency band from 863 to 870 MHz in parts of Europe.

Realizing this drawback in spectrum usage, Cognitive Radio (CR) was introduced to overcome this. CR allows new technologies to use the under-utilized spectrum in an opportunistic fashion, and it is defined as a radio that changes its transmission parameters given readings from the surrounding environment [18]. Hence, the CR

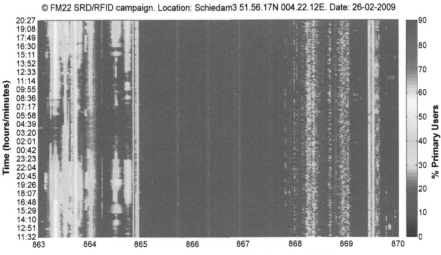

Figure 4. Spectrum utilization for 863 to 870 MHz band [17].

technology has two main responsibilities, the first is the cognitive capability, and the other one is the skill to reconfigure its hardware [16].

The cognitive ability entails monitoring of the surrounding environment, the use of the spectrum, and monitoring the power of the signals being transmitted by it, etc. On the other hand the ability to reconfigure refers to the ability of the CR technology to reconfigure and adapt to the parameters that will allow the exploitation and the usage of the under utilized spectrum.

CR introduces two types of users, i.e., (i) primary, and (ii) secondary users. The primary user is the one who was statically assigned the channel or the spectrum in question, while the secondary user is the one who will exploit and use this spectrum when the primary user is underutilizing it. As illustrated in Figure 5, the CR cycle starts with detecting the under-utilized space, followed by selecting the best frequency band to use, coordinating the spectrum access with other users, and finally, adjusting as per the situation. Adapting to the environment does not just include detecting a vacancy, but also being able to vacate the spectrum if the primary user is going to use it [19]. Hence, the functions that the CR utilizes to fulfill its responsibility can be divided into (i) spectrum sensing and analysis, and (ii) spectrum allocation and sharing.

As mentioned earlier, the CR senses the spectrum and analyzes it in order to decide when the secondary user can use to communicate, as the primary user is absent. Spectrum sensing techniques can be presented in three different categories, i.e., (i) the *interference temperature model*, (ii) the *spectrum-hole detection*, and (iii) the *cooperative sensing* [19].

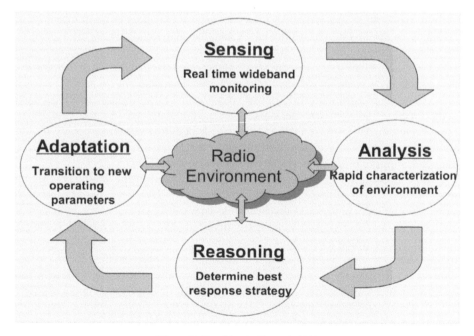

Figure 5. Cognitive Radio cycle [19].

The *interference temperature* can be defined as a parameter that measures the Radio Frequency (RF) power available at the receiving antenna to accommodate new signals [20]. The RF power available takes into account the amount of RF power received from noise, and other transmitters. In order to utilize this measure, an interference temperature limit is introduced as the threshold where interference from other sources can be tolerated. However, the practicality of this model has been questioned, and in May 2007, the FCC "rescinded its notice of proposed rule-making implementing the interference temperature model" [21].

When it comes to *spectrum-hole detection*, the techniques are categorized as one of the following: energy detector, feature detector, or matched filtering. In the energy detector technique, the sender checks for energy of received signal samples and compares it to a threshold value in order to decide whether a primary user is occupying the spectrum or not. This technique is easy to implement, and it does not require prior knowledge about the primary user's signals. However, a few disadvantages of this technique are that the threshold is highly susceptible to the variations in the level of noise. Another drawback is that this technique is not able to differentiate between the energy of a primary user versus that from a noise source or from interference. Moreover, the energy detector technique is not applicable when the direct sequence or the frequency hopping spread spectrum techniques are used by the transmitters [22].

On the other hand, the feature detector technique is based on coupling the modulated received signals with sine wave carrier, hopping sequence or other features of the primary signal characteristics [23]. This technique is better than the energy detector technique in the sense that it is more robust to the unknown noise levels, and can differentiate between the primary user signal and other noise or interference. However, some of the major drawbacks are that this technique requires previous knowledge of the primary user's signal characteristics, and it can only detect one signal at a given time. Finally, one last drawback is that it requires extensive processing and hence introduces some time delay in the spectrum detection.

The last type is the matched filtering which tests for the projected received signal in the direction of the primary user signal. This method is optimal because it maximizes the received signal-to-noise ratio (SNR), and hence, it is robust to noise. Another advantage is that it requires fewer signal sampling. However, it has high complexity and it also require prior knowledge of the primary user signal.

More spectrum sensing techniques were mentioned in the literature over the past few years. One of these methods is the statistical covariance based sensing, in which the statically covariance matrices of a received signal and noise are different, and hence, a received signal can be easily detected [24]. Another method is the fast sensing [25], in which a statistical test is done to check for change in the spectrum usage. Measurements-based sensing and modeling is another method in which long term spectrum monitoring is conducted to uniquely analyze the usage of the spectrum by primary users, and hence is used to develop dynamic access protocols [26], [19].

However, previously mentioned spectrum-sensing techniques do not provide a complete picture and precise information of the status of the spectrum and the presence of a primary user. This is mainly attributed to the uncertainty that arises from the multipath fading, shadowing and path loss [26]. In order to overcome these problems and provide a more efficient spectrum-sensing mechanism, *cooperative sensing* has been proposed. As illustrated in Figure 6, cooperative sensing can help improve and overcome the loss due to multipath and shadowing [28].

In cooperative sensing, different users sense the spectrum and share the information with each other. This sharing can occur either in a centralized, distributed, or in a relay-assisted fashion. In centralized sensing, a user is assigned the responsibility of selecting one specific channel to sense, and then it informs the users in the network to begin sensing and sending the information they retrieve back to it. Finally, it makes a decision about the presence of primary users in that spectrum and disseminates that information to the rest of the users in the CR network.

On the other hand, in the distributed approach, all users sense the spectrum and they reach a common decision about the presence or the absence of the primary user. In the relay assisted approach, some users who experience weaker sensing signals, can forward the information and act as a relay between users with a better sensing ability or channel.

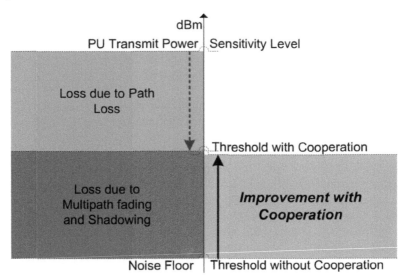

Figure 6. Improvement caused by the use of cooperative sensing [29].

Spectrum Allocation and Sharing

Following the spectrum sensing, the CR secondary user has to be able to adapt quickly to the presence or absence of the primary user. At any instance in time, the primary user may start using the spectrum, or other secondary users may start competing for it. Hence, the secondary users have to follow a spectrum allocation and sharing technique that will govern which spectrums will they sense, when will they access it, how to coexist with primary users, when to change the channel to another primary user and other related issues.

A few classifications can be used to differentiate between the various spectrum allocation and sharing techniques. One classification is according to what spectrum is being used, where some techniques are for accessing the unlicensed ISM band only, while other are for the licensed band. Another classification considers the access technology for the licensed spectrum, and can be categorized as either spectrum underlay, or spectrum overlay. In the first category, the secondary user is allowed to use the spectrum even in the presence of the primary user as long as the interference temperature metric at the primary users' receiver is below the acceptable threshold. In this case, the secondary user usually utilizes the spread spectrum techniques in order to keep the signal lower than the interference temperature limit. This is opposite to spectrum overlay, in which the secondary user is not allowed to use the spectrum in the presence of a primary user.

The last two classifications depend on the network architecture and the access behavior. When it comes to network architecture, some schemes are centralized where a central entity is responsible for allowing the access and coordinating the spectrum. Other schemes are distributed in which each user is responsible for making their own decision on spectrum access. When it comes to access behavior, some schemes are built around cooperation between the different users in the CR network and work towards a common goal, while others are non-cooperative and each user works for his interest and towards his own objective [19].

3. Cognitive Vehicular Networks

As mentioned earlier, there were a few motivations for using cognitive radio for vehicular network communications. Besides what has been listed, Cognitive Radio Vehicles (CRV) enhance some of the applications for VANETs. For example, when it comes to V2V communication and in case of safety applications or other information, the traditional approach uses the 5.9 GHz band. However, this band allows for a limited distance communication, while if lower frequencies are used, the signal will travel for a further distance. Another example is the case of public safety communication in case of disasters: the use of traditional infrastructure is not feasible, and the use of cognitive radio to access the unlicensed band would be possible [30].

However, the CRV has some different characteristics and features that are unique to vehicular networks compared to traditional cognitive radio. One of these characteristics is the impact of the mobility of the vehicles, which present a disadvantage since the spectrum sensed at one point in time has to keep in consideration the vehicle speed and the presence of different obstacles and surroundings.

Some other characteristics are advantages such as the fact that the vehicles follow a defined pattern of roads, and hence, if they communicate cooperatively with other vehicles ahead of them, they can gain knowledge of the available spectrums before physically reaching that point. One other advantage for CRVs is that the IEEE 802.11p standard defined a Common Control Channel (CCC). This channel can be utilized for coordination in some of the techniques used in CR.

Existing CRV Schemes

Some work in research presents schemes specifically designed for CRVs. Figure 7 depicts the breakdown of these scheme categories that are focused on two main areas,

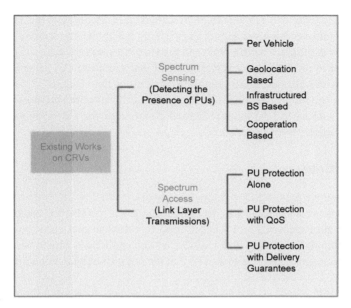

Figure 7. Existing work on CRVs classification [30].

i.e., (i) *spectrum sensing*, and (ii) *spectrum access*. In spectrum sensing, one approach is in which the vehicles can sense the spectrum autonomously, and make their own decisions accordingly.

A different approach would be the existence of a geographical database that includes information about the primary users and the spectrums available in each geo-location. The secondary user will only have to retrieve information from that database to simplify the spectrum analysis process. In infrastructure Base Station (BS) based sensing, the secondary user retrieves information from an infrastructure instead of a geo-location database. Finally, in cooperation based sensing, the secondary user cooperates with other users in order to analyze the spectrum.

With regard to spectrum access, the schemes can be classified into either access with focus on PU protection alone from secondary user signal interference, or with focus also on the quality of service for the vehicular application as well as the PU protection, or with focus on the delivery guarantee of the packets in the network.

CRV Open Research Issues

Some issues remain open in research when it comes to CRVs. One of them is the issue of impact of the mobility of the vehicle on the management of the spectrum. As discussed in [30], some of the open questions regarding the mobility are the need to study the effect of the vehicle mobility on the sensing of the spectrum, and whether the mobility predictability of the vehicles can be exploited to increase the spectrum awareness. Another issue is the optimal technique to reach a balance between the cooperation and spectrum scheduling for mobile vehicles using CR.

Another aspect that remains to be investigated in literature is the security of the CRVs. In cognitive radio, security is a main issue due to the cooperative nature of the network. For example, a selfish node can provide information about the presence of a primary user in order to have exclusive use of that spectrum. All these issues remain a problem with CRVs especially with the mobility of the vehicles.

Finally, the mobility and simulation of the CRV networks remains to be a main challenge. Authors in [31] have presented a simulator for CRVs; however, a more accurate simulator is still needed.

4. Conclusions

In this chapter we presented the vehicular network technology, its importance, and the different applications that can arise from its deployment. We also presented the different technology advancements and what is used to establish and carry communication between vehicles. Following that, main issues and challenges with these technologies have been presented, and it is observed that the cognitive radio technology can bring some relief.

The cognitive radio technology has been presented, its steps, approaches and the different schemes used. Finally, cognitive vehicular networks have been discussed, and the differences between them and traditional cognitive radio networks have been identified. A summary of the existing CRV schemes has been introduced, and some of the open research issues in CRV were discussed.

In conclusion, CRV offers a great promise for the improvement of vehicular networks, and can overcome many of the challenges observed in traditional VANETs.

References

[1] A. Mostafa. Packet Delivery Delay and Throughput Optimization for Vehicular Networks. Electronic Thesis or Dissertation, University of Cincinnati, 2013, Ohio LINK Electronic Theses and Dissertations Center, 23 Feb. 2015.

[2] C. Cordeiro and D.P. Agrawal. Ad hoc & Sensor Networks, Theory and Applications, 2nd edition, World Scientific Publishing, Spring, 2011.

[3] F.D. da Cunha, A. Boukerche, L. Villas, A. Carneiro Viana and A.A.F. Loureiro. Data Communication in VANETs: A Survey, Challenges and Applications [Research Report] RR-8498, INRIA Saclay, 2014. https://hal.inria.fr/hal-00981126v2.

[4] N.H.T.S. Admin. Vehicle Safety Communications-Applications (VSC-A) Project, Final Report, 2006–2011. file:///C:/Users/dpa/Downloads/811492A.pdf.

[5] Customers Angered as iPhones Overload ATT. http://www.nytimes.com/ 2009/09/03/technology/companies/03att.html.

[6] F. Borgonovo, A. Capone, M. Cesana and L. Fratta. Adhoc mac: New mac architecture for ad hoc networks providing efficient and reliable point-to-point and broadcast services. Wireless Networks, July 2004, 10(4): 359–366.

[7] Y.-B. Ko, V. Shankarkumar and N.H. Vaidya. Medium access control protocols using directional antennas in ad hoc networks, in Proc. of the 19th Annual Joint Conference of the IEEE Computer and Communications Societies (INFOCOM'00), pp. 13–21, 2000.

[8] Partners for Advanced Transportation Technology. http://www.path.berkeley.edu.

[9] J. Liu and S. Singh. ATCP: TCP for mobile ad hoc networks, IEEE Journal on Selected Areas in Communications, 2001, 19(7): 1300–1315.

[10] B. Jarupan and E. Ekici. A survey of cross-layer design for VANETs Ad Hoc Networks, July 2011, 9(5): 966–983.

[11]   R. Schmilz, A. Leiggener, A. Festag, L. Eggert and W. Effelsberg. Analysis of path characteristics and transport protocol design in vehicular ad hoc networks, in Proc. of the IEEE 63rd Vehicular Technology Conference, May 2006 (VTC 2006-Spring), 2: 528–532.

[12]   M.D. Dikaiakos, S. Iqbal, T. Nadeemand and L. Iftode. VITP: An information transfer protocol for vehicular computing, in Proc. of the 2nd ACM International Workshop on Vehicular Ad Hoc Networks (VANET'05), 2005, pp. 30–39.

[13]   M. Gerla, B. Zhou, Y. Zhong Lee, F. Soldo, U. Lee and G. Marfia. Vehicular grid communications: the role of the internet infrastructure, in Proc. of the 1st International Conference on Genetic Algorithms and their Applications, pp. 112–120, 2006.

[14]   Vehicle-Infrastructure Integration (VII) Initiative Benefit-Cost Analysis: Pre-Testing Estimates. http://www.its.dot.gov/research_docs/pdf/ 8cost-analysis.pdf.

[15]   Data overload threatens mobile networks. http://www.ft.com/intl/cms/s/0/caeb0766-9635-11e1-a6a0-00144feab49a.html.

[16]   Spectrum Policy Task Force Report, 2002: FCC, Cambridge University Press, December 2012. http://ebooks.cambridge.org/chapter.jsf?bid=CBO9781139058599&cid=CBO9781139058599A006.

[17]   CEPT Workshop on Spectrum Occupancy Measurements. http://cept.org/ecc/tools-and-services/cept-workshops/cept-workshop-on-spectrum-occupancy-measurements.

[18]   S. Haykin. Cognitive radio: Brain-empowered wireless communications. IEEE Journal on Selected Areas of Communication, 2005, 23(2): 201–220.

[19]   Beibei Wang and K.J.R. Liu. Advances in cognitive radio networks: A survey. IEEE Journal of Selected Topics in Signal Processing, Feb. 2011, 5(1): 5–23.

[20]   P.J. Kolodzy. Interference temperature: A metric for dynamic spectrum utilization. International Journal of Network Management, 2006, 16(2): 103–113.

[21]   T.C. Clancy. Dynamic spectrum access using the interference temperature model. Annals of Telecommunications, Springer, 2009, 64(7): 573–592.

[22]   D. Cabric, S.M. Mishra and R.W. Brodersen. Implementation issues in spectrum sensing for cognitive radios. Conference Record of the Thirty-Eighth Asilomar Conference on Signals, Systems and Computers, 2004, 7–10 Nov. 2004, 1: 772–776.

[23]   M. Maurizio and P. Vlad. Cognitive Radio Communications for Vehicular Technology—Wavelet Applications, Vehicular Technologies: Increasing Connectivity. Dr. Miguel Almeida (Ed.), ISBN: 978-953-307-223-4, In: Tech, DOI: 10.5772/14913. Available from: http://www.intechopen.com/books/vehicular-technologies-increasing-connectivity/cognitive-radio-communications-for-vehicular-technology-wavelet-applications.

[24]   Y. Zeng and Y.C. Liang. Spectrum-sensing algorithms for cognitive radio based on statistical covariances. IEEE Transactions on Vehicular Technology, 2009, 58(4): 1804–1815.

[25]   H. Li, C. Li and H. Dai. Quickest spectrum sensing in cognitive radio. Proc. 42nd Annual Conference on Information Science and Systems (CISS'08), 2008, pp. 203–208.

[26]   D. Willkomm, S. Machiraju, J. Bolot and A. Wolisz. Primary users in cellular networks: A large-scale measurement study. Proc. IEEE DySPAN, 2008, pp. 1–11.

[27]   I.F. Akyildiz, Won-Yeol Lee, C. Mehmet Vuran and Shantidev Mohanty. NeXt generation/dynamic spectrum access/cognitive radio wireless networks: a survey. Computer Networks Vol. 50, 13 September 2006, pp. 2127–2159.

[28]   S. Mishra, A. Sahai and R. Brodersen. Cooperative sensing among cognitive radios. Proc. of IEEE ICC 2006, 4: 1658–1663.

[29]   I.F. Akyildiz, B.F. Lo and R. Balakrishnan. Cooperative spectrum sensing in cognitive radio networks: a survey. Physical Communications, 2011, 4(1): 40–62, doi:10.1016/j. phycom.2010.12.003.

[30]   M. Di Felice, R. Doost-Mohammady, K.R. Chowdhury and L. Bononi. Smart Radios for smart vehicles: cognitive vehicular networks. IEEE Vehicular Technology Magazine, June 2012, 7(2): 26–33.

[31]   M. Di Felice, K.R. Chowdhury and L. Bononi. Cooperative spectrum management in cognitive Vehicular Ad Hoc Networks. IEEE Vehicular Networking Conference (VNC), Nov. 14–16, 2011, Vol., no.: 47–54.

<u>Part II</u>

Cognitive Radio for Vehicular Networks

3

Policy-Based Distributed Spectrum Allocation

Flavio Esposito[1],* and Donato Di Paola[2]

ABSTRACT

Cognitive radios are a disruptive technology based on the fundamental software-defined mechanism that enables wireless devices to sense their environment, and dynamically adapt the transmission method on the currently available spectrum. When such wireless devices are embedded in vehicles, the single-node challenges of dynamic radio resource allocation are exacerbated by the dynamic nature of the network. A vehicular network may have in fact variable contact rates, variable hop-to-hop and end-to-end delays, variable traffic density, and even variable goals, given by each service or application. In such complex heterogeneous and dynamic scenarios, a "one-size fits all" spectrum allocation solution possibly cannot exist.

To this end, in this chapter we propose two policy-based solutions that allow cognitive vehicular networks to distributively adapt locally to the available spectrum and globally to a service goal or objective by merely instantiating a few policies. In particular, in the first part of the chapter we model the spectrum allocation as a global utility maximization problem. We then show a policy-based technique that leverages decomposition theory to tune the behavior of the multi-dimensional spectrum allocation solution. In the second part of the chapter we focus on another policy-based technique. Leveraging recent results on the consensus literature we propose an approach that enables cognitive radio-equipped vehicles to allocate the spectrum in a distributed fashion, providing guarantees on both convergence time and performance with respect to a Pareto optimal spectrum allocation.

[1] Computer Science & IT Department, University of Missouri, Columbia.
[2] Institute of Intelligent Systems for Automation, National Research Council (ISSIA-CNR), Bari, Italy.
* Corresponding author: espositof@missouri.edu

1. Introduction

Cognitive Radio Networks (CRNs) are a novel communication paradigm whose goal is to efficiently and effectively utilize the radio spectrum by dynamically sensing and accessing the available and temporarily unused frequencies. Cognitive radio equipped vehicles sense the spectrum of the licensed users, also known as *Primary Users* and opportunistically utilize vacant sub-bands (also known as spectrum holes or white space) within the Primary Users licensed spectrum. Cognitive Radio Networks users are often denoted in the literature as *Secondary Users* as they opportunistically use the temporarily unused and regulated spectrum of the *Primary Users* [15].

In this promising technology, which addresses the well-known problem of allocating scarce but low-utilized spectrum resources, four separate spectrum management mechanisms have been identified [1]:

1. **Spectrum Sensing:** vehicles determine which portions of the spectrum is available.
2. **Spectrum Selection:** vehicles select the best available channel according to some network or vehicle utility function.
3. **Spectrum Sharing:** vehicles coordinate with other vehicles the access to the available channel.
4. **Spectrum Mobility:** vehicles clear the channel when a licensed user is detected.

The focus of this chapter is on a particular vehicle spectrum management sub-problem which deals solely with selecting the appropriate available frequency set in a distributed fashion for cognitive vehicular network applications.

The spectrum allocation problem in CRNs is more challenging than in other existing wireless communication networks for several reasons. First, the required detection of errors in spectrum sensing algorithms deteriorates the performance of the adapting spectrum allocation algorithm [18, 31]. Second, cognitive vehicular networks are often required to be self-regulating, so centralized management approaches are often inappropriate to control, e.g., interference [7, 26, 31]. Third, due to the absence of a licensed spectrum, the available wireless resources are usually temporary and unstable, which results in a higher probability of transmission interruptions. One partial solution to the latter problem is to enable vehicles to search for spectrum resources from multiple primary networks simultaneously, resulting though into a more complex resource allocation problem. In the second part of the chapter (Section 4) we show a spectrum allocation mechanism in which vehicles are allowed to simultaneously request multiple frequencies using a distributed consensus-based auction.

Due to the potentially different conditions and constraints present in environments surrounding a cognitive vehicular network, a one-size-fits-all spectrum allocation solution possibly cannot exist. The same spectrum allocation mechanism should be instantiated with different policies to adapt to different vehicles' or applications' goals.[1] Examples of conditions that can dictate the policies for a (distributed) spectrum allocation mechanism in cognitive vehicular networks are:

- **Exogenous and Application Conditions:** Different speed of cognitive radio-equipped vehicles, or different vehicles traffic density may lead to different convergence time to an optimal spectrum allocation. For example, in a scenario with high density traffic, where the topology does not change at a very high rate, an iterative algorithm has enough time to reach an optimal allocation; in low-density traffic or high-speed scenarios, a distributed iterative algorithm has to reach instead a stopping condition much faster, before the connectivity with other vehicles is lost.

- **Channel Conditions:** Cognitive vehicular networks with high delay require distributed solutions to converge faster, while high bandwidth links and low noise level networks consent a higher number of network state exchanges. Moreover, different available channels may require vehicles to clear the resource allocation problem within a single channel or across multiple available channels.

- **Distribution Model:** Applications may require (i) a centralized agent computing the global optimal allocation, (ii) a fully distributed ad-hoc network where no central authority has a full state knowledge of the entire vehicular network, or (iii) a decentralized third party agent, e.g., a telecommunication tower which collects the costs of allocating a given frequency on each vehicle and updates such vehicles with the current optimal spectrum allocation, or decides the transmission rates for each channel based on the speed of the vehicles [10].

Chapter Contributions and Organization: In the first part of this chapter (Section 2) we define the cognitive VANET spectrum management as a centralized optimization problem, capturing its network utility maximization nature. The model is general enough to be adapted to different cognitive vehicular network goals and conditions. Moreover, the model serves as a general architectural framework to allow cognitive vehicular network applications to adapt the mechanisms to several goals (utility) and scenarios (constraints) by merely instantiating different policies without having to reimplement ad-hoc distributed allocation algorithms.

Due to the rich structure of the optimization problem, many ways to decompose the problem exist. We leverage standard decomposition techniques to give a few examples of how the spectrum allocation mechanisms can be reprogrammed with different policies (Section 3). An example of policy is the type of decomposition: primal or dual. Each particular multi-level combination of primal/dual decompositions provides a different distributed algorithm for the spectrum allocation network utility maximization problem. The choice of the adequate decomposition method and distributed algorithm depends on factors such as the amount of signaling required for

---

[1] By mechanisms, we mean the invariant aspect of a problem. By policy, we mean a variant aspect of a mechanism, i.e., a user goal or condition. *Acknowledgment* is an example of a mechanism, while *when to acknowledge* is an example of policy. It may be best to modify the acknowledgment rate to adapt to different traffic conditions or interference levels.

proper coordination, the robustness, and the speed of convergence. In Section 3.3 we show few simulation results depicting a representative tradeoff analysis that emerges from the instantiation of the cognitive network spectrum allocation problem with either primal or dual decomposition policies. The example clarifies how this engineering tradeoff is regulated by the amount of variables as well as the number of constraints that characterize the specific spectrum allocation problem.

In the second part of the chapter (Section 4), we leverage instead *max-consensus auctions* [8, 24] to propose a different policy-based technique to adapt the spectrum allocation problem to different cognitive vehicular network applications' goals. In particular, we describe the design of a distributed auction which allows vehicles to independently bid on the available spectrum resources in a cooperative manner to reach a *Pareto optimality* (maximize the summation of all vehicles' utilities), followed by an agreement phase where the auction is cleared in a distributed fashion. The idea of using max-consensus auction as a distributed allocation problem was borrowed by our earlier work on distributed virtual network embedding [8], but it has applications in a wider range of resource allocation problems [5, 9, 24].

Even though this chapter focuses on the spectrum allocation mechanism, our model can be extended to capture the architectural interactions between spectrum allocation and other fundamental cognitive radio network mechanisms by properly adjusting the utility function and by removing or adding constraints. Any spectrum management task may be either architecturally isolated, i.e., solved in isolation within a modular software or hardware component, or jointly solved with other functionalities, i.e., in a cross-layer or cross-functional optimization.

2. The Spectrum Allocation Problem for Cognitive Vehicular Networks

Having motivated the need for a policy-based, flexible spectrum allocation solution in Section 1, in this section we formally define the *Cognitive Vehicular Spectrum Allocation Problem* as a multi-dimensional allocation problem, and then we use optimization theory to define a mixed integer programming problem that models its behavior.

Consider a network of n agents to which m objects have to be allocated.[2] Let us indicate with $\mathcal{I} = \{1,...,n\}$ and $\mathcal{J} = \{1,...,m\}$ the index sets of agents and objects, respectively. In a multi-allocation setting, it is assumed that an agent $i \in \mathcal{I}$ can be assigned to more than one object. Thus, each agent i is associated to a capacity $l \in \mathbb{N}$ which indicates the maximum number of objects that can be allocated to an agent. An important objective of any multi-assignment problems is to obtain a conflict-free

---

[2] Later in the chapter we will refer such objects as frequency of the vehicular cognitive network, but the problem is general enough to consider objects. Such objects can represent any spatio-temporal resource to be allocated. For example, the spectrum may be divided among vehicles using time division multiplexing, not only frequency division multiplexing. Similarly, even though later in the chapter we refer to vehicles, in the problem formulation we use the term agents, to indicate not only vehicles, but telecommunication towers, applications or service providers, or any other entity participating to the cognitive vehicular network state exchange, as long as they have an (IP) address and computational power.

assignment. An assignment is called *free of conflicts* if each object is allocate to one and only one agent in the network. Furthermore, we assume that the network is composed by heterogeneous agents, i.e., agents with different capabilities, and we define a matrix of possible allowed assignments $\Delta \in \{0,1\}^{n \times m}$. Each element of the matrix Δ_{ij} is equal to one if and only if agent i is allowed to allocate (or host) the object identified with j, and zero otherwise. To find the optimal assignment, each agent is associated to a utility value $U_{ij} \in \mathbb{R}_+$, which intuitively indicates a non-negative return obtained by agent i when hosting object j. Given these assumptions and notations, we formulate the *Cognitive Vehicular Spectrum Allocation Problem* as follows:

Problem 2.1 (The Cognitive Vehicular Spectrum Allocation Problem) *Given a network of n heterogeneous agents, e.g., vehicles and m objects, e.g., frequencies, the Cognitive Vehicular Spectrum Allocation Problem is the problem of finding the optimal object allocation, subject to the following set of constraints: (i) all data transmissions need to be interference-free (conflict free allocation), (ii) we cannot assign more objects (time slots or frequencies) than each vehicle's radio can support, and (iii) all the data transfer demands among vehicles needs to be met, i.e., the multi-commodity flow constraints need to be satisfied.*

To model the constraints imposing that all vehicle's data transfer demands need to be met, we use the standard multi-commodity flow [11]. The multi-commodity flow problem takes into account the transferring of multiple commodities (flow demands) of data among multiple agents, given a set of bandwidth capacity. In the following, we denote the demand with the index k, element of the index set $K = \{1,...,p\}$.

Formally, we model the *Cognitive Vehicular Network Spectrum Assignment Problem* with the following optimization problem, with decision variables z_{ij} and f_k:

$$\max_{z, f_k} \sum_{i=1}^{n} \sum_{j=1}^{m} U_{ij}(z_{ij}, f_k) \tag{1}$$

subject to:

```
capacity constraints
```

$$\sum_{j=1}^{m} z_{ij} \leq l \quad \forall i \in \mathcal{I} \tag{2}$$

```
interference-free constraints
```

$$z_{ij} \leq \Delta_{ij} \quad \forall (i, j) \in \mathcal{I} \times \mathcal{J} \tag{3}$$

```
multi-commodity flow constraints
```

$$\sum_{k=1}^{K} f_k(u, v) \leq \mathcal{R}(u, v) \quad \forall (u, v) \in \mathcal{I} \tag{4}$$

$$\sum_{w \in W} f_k(u, w) = 0 \quad \forall k \ \forall \ u \neq s_i, t_i \tag{5}$$

$$f_k(u, v) = -f_k(v, u) \quad \forall k \ \forall \ (u, v) \in \mathcal{I} \tag{6}$$

$$\sum_{w \in W} f_k(s_k, w) = \sum_{w \in W} f_l(w, t_k) = d_k \quad \forall s_k, t_k \tag{7}$$

existential constraints

$$f_k \in [0, 1] \quad \forall \, k \in K \tag{8}$$

$$z_{ij} \in \{0, 1\} \quad \forall (i, j) \in \mathcal{I} \times \mathcal{J} \tag{9}$$

where $z_{ij} = 1$ if frequency $j \in \mathcal{J}$ is assigned to vehicle $i \in \mathcal{I}$ and 0 otherwise, and $U_{ij} \geq 0$ is the utility function. Constraints (2)-(3) express, respectively, the requirement that the number of frequencies assigned to each vehicle should not exceed its capacity l (number of channels), and that a frequency should be assigned to no more than one vehicle in the wireless neighborhood, to avoid interference. An element on the assignment matrix is denoted with 1 if a vehicle is equipped with the radio technology operating on frequency j, and 0 otherwise. Constraints (4)-(7) are the standard flow conservation constraints. In particular, constraints (4) ensure that the requested capacity of each flow is lower than the capacity $\mathcal{R}(u, v)$ of the channel connecting agent u with agent v. Constraints (5) ensure that the net flow on each physical link is zero, i.e., every outgoing bit from a vehicle transmitter s_k is received by the destination vehicle t_k and not dispersed within any other vehicle w in the vehicle neighborhood W. Constraints 6 ensure that each outgoing flow from a source u to a destination v is equivalent to the incoming flow to v from vehicle u, and finally, constraints 7 ensure that all the data transferring demands d_k are met.

Remark 1 *Note how any spectrum allocation problem runs under the assumption that (i) at least an initial vehicle's topology has already been discovered by a standard network discovery protocol, and (ii) an initial range of available (time slots or) frequencies is available, in order to enable communication for the execution of any distributed algorithm. We believe that an interesting research direction is to design and implement distributed heuristics or approximation algorithms that simultaneously optimize the objectives of both network discovery and spectrum allocation, as well as managing radio interference.*

In the next subsection we show with a proof sketch that the spectrum allocation problem in cognitive vehicular networks is NP-hard, even for two objects (e.g., frequencies) and two agents (e.g., vehicles), motivating the investigation of polynomial time heuristics to approximate it.

2.1 Problem Complexity

Being a combination of other NP-hard problems, it is trivial to show that the cognitive vehicular network spectrum allocation problem is NP-hard. We omit the formal proof but we give a general intuition. Let us consider the multi-commodity flow constraints of the problem. In its decision version, the problem of producing an integer flow satisfying all demands has been shown to be NP-complete [11], even when only two flow commodities and their capacities are unitary, making the problem strongly NP-complete in this case. If fractional flows are allowed, i.e., if a flow can be split among multiple "virtual channels" among vehicles, the problem can be solved in polynomial time through linear programming [6] or via fully polynomial time approximation schemes [19], typically much faster. Aside from the multi-commodity flow constraints, we note how few of the other constraints are equivalent to the set packing problem (known to be an NP-hard problem [16]). An optimal algorithm able to determine the capacity and the interference free constraints would have to try all possible combinations of assignments to decide which is the candidate that maximizes the network utility.

3. Decomposition-based Distributed Spectrum Allocation

As we have seen in the previous subsection, the problem of finding the optimal frequency allocation that maximizes the network utility in a vehicle network is *NP-hard*. The challenges of solving such problem are exacerbated by the distributed nature of vehicular networks. As detailed in Section 1, it is in fact impractical for a single centralized entity to clear the frequency allocation problem. Moreover, different distributed scenarios and vehicular network applications may have different goals; a *one-size fits all* distributed solution for the cognitive spectrum allocation problem probably cannot exist. To this end, in Sections 3.1 to 3.3 we use decomposition theory to dissect the problem of a distributed spectrum allocation. The idea of using decomposition theory in the field of cognitive radio has been floated before. For example, the authors of [31] presented a primal-dual decomposition-based cross-layer scheduling for power allocation and sub-channel assignment. In this chapter we do not use decomposition techniques to propose a solution to solve a specific problem, but we model a general architectural framework which can be used to reprogram many spectrum allocation solutions, by merely tweaking a set of policies that modify the utility function and the problem constraints.

3.1 Decomposition Architecture

Due to the rich structure of Problem (1), many different decompositions are possible. Each alternative decomposition leads to a different distributed algorithm, with potentially different desirable properties. The choice of the adequate decomposition method and distributed algorithm for a particular problem depends on the vehicles' goals, and on the offered service or application. The idea of decomposing Problem (1) is to convert it into equivalent formulations, where a master problem interacts with a set of subproblems. Decomposition techniques can be classified into *primal* and *dual*. Primal decompositions are based on decomposing the original primal Problem (1), while dual decomposition methods are based on decomposing its dual [4].

In a primal decomposition, the master problem allocates the existing resources by directly assigning to each subproblem the amount of resources that it can use. Dual decomposition methods instead correspond to a resource allocation via pricing, i.e., the master problem sets the resource price for all the subproblems, that independently decide if they should transmit using a given frequency or not, based on such prices.

Primal decompositions are applicable to problem (1) by an iterative *partitioning* of the decision variables into multiple subsets. Each partition set is optimized separately, while the remaining variables are fixed. For example, in a vehicular network using the wi-fi technology, and following the IEEE 802.11 b/g/n standards, we could first optimize the allocation of frequencies for channel 1 and 6, fixing the remaining channel 11, and then optimize the allocation of channel 11 in a second phase, given the optimal value of the first set of decision variables.[3] Alternatively, a distributed dynamic frequency allocation algorithm could simultaneously optimize all three channels for geographically distributed partitions of the wireless vehicular network. Primal decompositions can also be applied with respect to the two subproblems: frequency assignment (nodes) and multi-commodity flow (links). For example, by fixing the multi-commodity flow variables f_k, the remaining subproblem can be solved by optimizing the frequency allocation variables first, and then optimize the variables z_j once the optimal variables $\mathbf{f}_k^\star$ are known.

Dual decomposition approaches are based on decomposing the Lagrangian function formed augmenting the master problem with the relaxed constraints. Even in this case, it is possible to obtain different decompositions by relaxing different sets of constraints, hence obtaining different distributed algorithms. Regardless of the number of constraints that are relaxed, dual decompositions are different than primal in the amount of required parallel computation (all the subproblems could be solved in parallel), and by the amount of message passing between one phase and the other of the iterative method. The dual master problem communicates to each subproblem the shadow prices, i.e., the Lagrangian multipliers, then each of the subproblems (sequentially or in parallel) is solved, and the optimal value is returned, together with the subgradients. It is also possible to devise solutions in which both primal and dual decompositions are used.

In general, a centralized service provider or a distributed application can instantiate a set of policies at the *master problem*, dictating the order in which the variables need to be optimized among the vehicles, and on which (wireless network) partition. The subproblems resulting from the decomposition can also instantiate other sets of decomposition policies, to decide which variables are to be optimized next, in which order, or even further decomposing the subproblems (Figure 1). In Section 3.2 we show two examples of such policy instantiation, solving the distributed spectrum allocation problem with iterative methods resulting from a primal and a dual decomposition of the original (master) problem.

---

[3] Note how in IEEE 802.11 b/g/n only channels 1, 6 and 11 can be used simultaneously (not only be a cognitive network) to avoid interference.

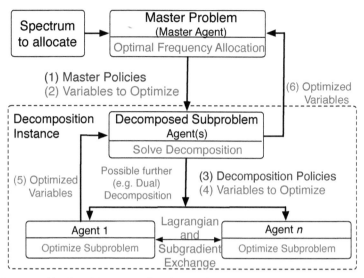

Figure 1. Decomposition-based Spectrum Allocation Architecture: different spectrum allocation solutions can be modeled via primal and dual decompositions. A multi-radio equipped agent $i \in \mathcal{I}$ instantiates via its configuration a problem formulation according to its policies (1), and picks an objective function U_i (2). The other agents solve the decomposed subproblems, possibly further decomposing them (3-4). Finally, the optimal spectrum allocation variables are returned to the main master agent (5-6), that eventually releases the next frequency to allocate.

3.2 Primal versus Dual Decompositions

In this section we analyze the tradeoffs between primal and dual decompositions. We consider problem (1) reduced to a normal form, where the decision variables have been split in two partitions:

$$\max_{p,q} \quad c^T p + \tilde{c}^T q$$

subject to $Ap \leq b$ (10a)

$$\tilde{A}q \leq \tilde{b} \tag{10b}$$

$$Fp + \tilde{F}q \leq h \tag{10c}$$

where p and q are the sets of decision variables referring to the first and to the second vehicular network partition, respectively; F and $\tilde{F}$ are the matrices of constraints for the two partitions, and h is the vector of all capacity limits. The constraints (10a) and (10b) capture the separable nature of the problem. Constraint (10c) captures the complicating constraint.

Procedure 1 Distributed Spectrum Allocation by Primal Decomposition

---

1: Given ζ at iteration t, solve subproblems to obtain optimal spectrum allocation ϕ and $\tilde{\phi}$
   for each partition, and dual variables $\lambda^\star(\zeta_t)$ and $\tilde{\lambda}^\star(\zeta_t)$
2: Send/Receive ϕ, $\tilde{\phi}$, $\lambda^\star$ and $\tilde{\lambda}^\star$
3: Master computes subgradient $g(\zeta_t) = -\lambda^\star(\zeta_t) + \tilde{\lambda}^\star(\zeta_t)$
4: Master updates resource vector $\zeta_{t+1} = \zeta_t - \alpha_t g$

Primal Decomposition. By applying primal decomposition to problem (10), we can separately solve two subproblems, one for each partition, by introducing an auxiliary variable ζ, which represents the percentage of frequencies and flows allocated to each subproblem. The original problem (10) is equivalent to the following master problem:

$$\max_{\zeta} \; \phi(\zeta) + \tilde{\phi}(\zeta) \tag{11}$$

where:

$$\phi(\zeta) = \quad \sup_p \; c^T p \tag{12a}$$
$$\text{subject to} \quad Ap \leq b \tag{12b}$$
$$\qquad\qquad\quad Fp \leq \zeta \tag{12c}$$

and

$$\tilde{\phi}(\zeta) = \quad \sup_q \; \tilde{c}^T q \tag{13a}$$
$$\text{subject to} \quad \tilde{A}q \leq \tilde{b} \tag{13b}$$
$$\qquad\qquad\quad \tilde{F}q \leq h - \zeta. \tag{13c}$$

The primal master problem (11) maximizes the sum of the optimal values of the two subproblems, over the auxiliary variable ζ. After ζ is fixed, the subproblems (12) and (13) are solved separately, sequentially or in parallel, depending on the application provider's policy or the policy of each vehicle. The master algorithm updates ζ and collects the two subgradients independently computed by the two subproblems. To find the optimal ζ we use a subgradient method [4]. In particular, to evaluate a subgradient of $\phi(\zeta)$ and $\tilde{\phi}(\zeta)$, we first find the optimal dual variables $\lambda^\star$ for the first subproblem subject to the constraint $Fp \leq \zeta$. Simultaneously (or sequentially) we find the optimal dual variables $\tilde{\lambda}^\star$ for the second subproblem, subject to the constraint $\tilde{F}q \leq h - \zeta$. The subgradient of the original master problem is therefore $g = -\lambda^\star(\zeta) + \tilde{\lambda}^\star(\zeta)$; that is, $g \in \partial(\phi(\zeta) + \tilde{\phi}(\zeta))$.[4] The primal decomposition algorithm, combined with the subgradient method for the master problem is repeated, using a diminishing step size, until a stopping criteria is reached (Procedure 1).

The optimal Lagrangian multiplier associated with the capacity of the vehicle i, $-\lambda_i^\star$, tells us how much worse the objective of the first subproblem would be, for a small (marginal) decrease in the capacity of the vehicle i. $\tilde{\lambda}_i^\star$ tells us how much better the objective of the second subproblem would be, for a small (marginal) increase in the capacity of vehicle i. Therefore, the primal subgradient $g(\zeta) = -\lambda(\zeta) + \tilde{\lambda}(\zeta)$ tells us how much better the total objective would be if we transfer some physical capacity of vehicle i from one subsystem to the other. At each step of the subgradient method,

---

[4] For the proof please refer to §5.6 of [4].

more capacity of each vehicle is allocated to the subproblem with the larger Lagrange multiplier (see Figure 3b). This is done with an update of the auxiliary variable ζ. The resource update $\zeta_{t+1} = \zeta_t - \alpha_t g$ can be interpreted as shifts of some of the capacity to the subsystem that can better use it for the global utility maximization.

Dual Decomposition. An alternative method to solve problem (10) is to use dual decomposition, relaxing the coupling capacity constraint (10c). From problem (10) we form the partial Lagrangian function:

Procedure 2 Distributed Spectrum Allocation by Dual Decomposition

---
1: Given λ_t at iteration t, solve the subproblems to obtain the optimal values $p^\star$ and $q^\star$ for each problem partition
2: Send/Receive optimal node embedding $p^\star$ and $q^\star$
3: Master computes the subgradient $g = F\,p^\star + \tilde{F}\,q^\star - h$
4: Master updates the prices $\lambda_{t+1} = (\lambda_t - \alpha_t g)_+$

$$L(p, q, \lambda) = c^T p + \tilde{c}^T q + \lambda^T (Fp + \tilde{F}q - h) \tag{14a}$$

Hence, the dual function is:

$$Q(\lambda) = \inf_{u,v}\{L(p, q, \lambda) | Ap \le b, \tilde{A}q \le \tilde{b}\} \tag{15a}$$

$$= -\lambda^T h + \inf_{Ap \le b} (F^T \lambda + c)^T p + \inf_{\tilde{A}q \le \tilde{b}} (\tilde{F}^T \lambda + \tilde{c})^T q,$$

and the dual problem is:

$$\max_{\lambda} \quad Q(\lambda) \tag{16a}$$

subject to $\lambda \ge 0$,

The Projected Subgradient Method. We solve problem (16) using the projected subgradient method [4]. To find a subgradient of Q at λ, we let $p^\star$ and $q^\star$ be the optimal solutions of the subproblems:

$$p^\star = \max_{p} \quad (F^T \lambda + c)^T p \tag{17a}$$

subject to $\quad Ap \le b \tag{17b}$

and

$$q^\star = \max_{q} \quad (\tilde{F}^T \lambda + \tilde{c})^T q \tag{18a}$$

subject to $\quad \tilde{A}q \le \tilde{b} \tag{18b}$

respectively. Then the vehicular network processes in charge of solving the subproblems send their optimal values to the master problem, so that the subgradient of the dual function can be computed as:

$$g = Fp^\star + \tilde{F}q^\star - h. \tag{19}$$

The subgradient method is run until a termination condition is satisfied (Procedure 2); the operator $(\cdot)_+$ denotes the non-negative part of a vector, i.e., the projection onto the non-negative orthant.

At each step, the master problem sets the prices for the frequencies to assign. The subgradient g, in this case, represents the margin of the original shared coupling constraint. If the subgradient associated with the capacity of vehicle i is positive $(g_i > 0)$, then it is possible for the two subsystems to use more capacity of the vehicle i. The master algorithm adjusts the price vector so that the price of each overused vehicle is increased, and the price of each underutilized vehicle is decreased, but never negative.

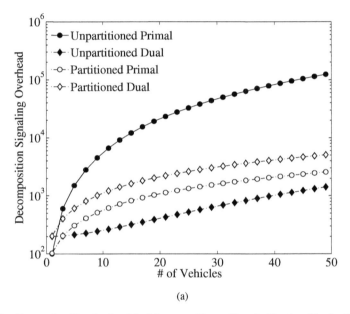

(a)

Figure 2. Signaling overhead in primal and dual decompositions with and without partitioning the spectrum decision variables into two sets of frequencies, each one optimized independently. As we can see, in one case the primal has lower overhead, while in the other case the dual has lower overhead. Note that the figure represents a single simulation run.

3.3 Primal vs Dual Decomposition Tradeoffs

A natural question to ask at this point is weather we should design a distributed cognitive vehicular network allocation algorithm that uses primal or dual decomposition. This engineering tradeoff is regulated by the amount of variables as well as the number of constraints in the problem. So there is no right answer, and this section will try to show it with a CPLEX-based simulated example. This implies that a careful design phase is required to optimize the resource allocation in cognitive vehicular networks.

Depending on the number of constraints and decisional variables, applying primal or dual decomposition may or may not decrease the overhead, i.e., the message passing among the master and the subproblems. Splitting the original problem into multiple subproblems causes more over-head between the master problem and the distributed subproblems, as more Lagrangian multipliers and intermediate sub-gradient values needs to be exchanged and updated. Moreover, an iterative algorithm obtained decomposing the original problem with a primal decomposition may or may not converge faster to an acceptable optimal solution.

In Figure 2 we analyze the tradeoff between the signaling overhead (number of exchanged messages between the master and the subproblems) as we increase the number of vehicles participating in the distributed resource allocation. We compare the solutions obtained using both primal and dual decomposition. In one case, we optimize all the decision variables simultaneously, while in the other case we partition the decision variables into two sets of equal size. As we can see, when we do not partition the variable sets, it is convenient to use a dual decomposition approach, but if we partition the variable sets, primal decomposition result into a lower overhead, even if the improvement is less pronounced. Weather or not we should partition the variable set depends on the application, and on the distribution model (see Section 1).

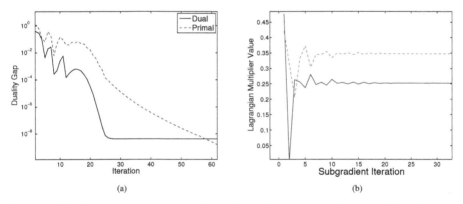

Figure 3. (a) Duality gap after decomposing the problem with primal or dual decomposition. After each decomposition, the subproblems are further decomposed with the dual method, and a sub-gradient algorithm is applied: primal decomposition results into a lower duality gap. (b) Convergence to the optimal Lagrangian multipliers when applying the iterative subgradient method.

Primal and dual decomposition may have different behavior even in terms of how fast they reach the optimality condition. In Figure 3a we plot the *duality gap* at each iteration of the subgradient algorithm, after decomposing the problem with primal and with dual decomposition. After each decomposition, the subproblems are solved with the dual method. As we can see, during the first 56 iterations, the duality gap is lower when the problem has been decomposed with primal decomposition. If instead we have a delay insensitive cognitive network scenario, i.e., an optimal frequency allocation is more important than a quicker suboptimal solution, than it is convenient to apply primal decomposition and stop after an higher number of iterations.

Note how, depending on the nature of the problem and the constraints (if the Slater's condition is not satisfied [4]), the optimality gap may not be reduced to zero at all (as in this case for the dual decomposition), so a careful design planning phase is even more important. Most of the time (just as in this case) the duality gap is extremely small (10^{-8}) so a suboptimal solution may be acceptable. To be convinced that few iterations may be enough in some cases, in Figure 3b we show another simulation scenario in which only two vehicles are competing for two frequencies. The evolution of the prices to be paid by each of the two subproblems (Lagrangian multipliers) does not change after merely 20 iterations.

4. Consensus-based Distributed Spectrum Allocation

In the previous section we have shown how the spectrum allocation problem can be solved in a distributed fashion using classical decomposition techniques, and how such solution can be adapted to different goals by merely instantiating few decomposition policies (Figure 1). In this section, we present an alternative approach to the distributed Spectrum Allocation problem. In particular, to obtain a fully distributed algorithmic framework, i.e., an environment in which all vehicles act as peers without supervision or a trigger from an external agent, we leverage the combination of auction-based algorithms and consensus theory [23, 28]. Auctions are an effective resource allocation mechanism when the resources' values are unknown to the bidders and to the auctioneer. In a distributed cognitive vehicular network, the range of available frequencies can be often at best only estimated, so auction-based approaches are a viable solution to our problem.

4.1 Distributed Max-Consensus Auctions Algorithms

To our knowledge, the first auction algorithm solving a resource allocation problem was proposed in 1988 by Bertsekas [2] as a polynomial-time algorithm to solve a single-assignment problem. The algorithm was centralized. Further modifications to the algorithm have been later proposed for multi-assignment problems and decentralized solutions.

In our settings, Bertsekas's algorithm would associate a profit $\{a_{ij} - p_{ij}\}$ to each vehicle $i \in \mathcal{I}$, seeking the maximum network utility. For each frequency j of the available radio spectrum $\mathcal{J}$, a_{ij} represents the value of such frequency to vehicle i, and p_{ij} the local cost (or price) that the vehicle has to pay to reserve such frequency. The cost is updated at every auction round by vehicle i to reflect the current locale value. The auction is performed until all available channels have been assigned, in a way which maximizes the summation of vehicles' profits, i.e., such that every vehicle i is assigned to the frequency j^*, where:

$$\pi_i = max_j(a_{ij} - p_{ij}). \tag{20}$$

If the frequency j^* has already been assigned to another vehicle's flow, the two vehicles swap frequency. Whenever there is a frequency swap, the new vehicle i now assigned to frequency j increases the price of j^* to $p_{ij\star}$, so that, in future auction rounds, only vehicles willing to pay a price grater than $p_{ij\star}$ are allowed to assign such frequency to themselves. This mechanism guarantees that every vehicle is assigned to the frequency giving the maximum value to the network utility, i.e., the auction seeks a *Pareto optimality*. In this class of algorithms, generally all the vehicles' bids are collected by an auctioneer [13, 21] to determine the winner based on the highest bid.

More recently, decentralized auction algorithms with new strategies to remove the need of an auctioneer have been proposed and developed; see, e.g., [29, 30, 32]. These algorithms introduced different conflict resolution approaches to determine the vehicle that "wins" the objects to allocate. An interesting solution to the decentralized multi-assignment problem, proposed to allocate tasks to a fleet of unmanned vehicles is the so called Consensus-Based Bundle Algorithm (CBBA) [5]. CBBA follows a greedy heuristic scheme that collects bids of agents on objects, and then resolves conflicts on the assignments a using max-consensus strategy among bids.

Consensus-based auction algorithms like CBBA [5] or CAD [8] are capable of enabling a (cognitive) network of (unmanned) vehicles to reach a distributed agreement on a state, e.g., a measurement, an action, an opinion, or in our case, a frequency or a time slot to be dynamically allocated, via merely local computations and first-hop neighbor communication, and with guarantees on the optimality of the assignment.

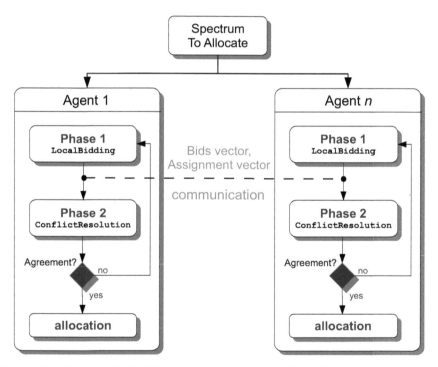

Figure 4. Max-Consensus Auction Workflow: agents independently bid on items, e.g., frequencies, that are later exchanged for a distributed winner determination.

Among different consensus approaches, max-consensus auctions have been shown to be a valuable tool for estimation and assignment problems [8, 9, 14, 27]. Given the vehicles communication network, usually represented by a graph $\mathcal{G} = (\mathcal{I}, \mathcal{E})$, where the vehicles constitute the node set $\mathcal{I}$ and the communication channels are modeled by the link set $\mathcal{E} \subseteq \mathcal{I} \times \mathcal{I}$, each vehicle i can update its local price for a given frequency j using the following (local) updating rule:

$$\tilde{p}_{ij}(t) = \max_{k \in \mathcal{N}^{[i]}} \{p_{kj}(t)\}, \tag{21}$$

where t denotes the iteration of the consensus algorithm, and $\mathcal{N}^{[i]}$ is the set of one-hop neighbors of vehicle i. The value $\tilde{p}_{ij}(t)$ is used by the vehicle i to determine on which frequency to bid on the next round. It has been shown that max-consensus algorithms converge towards an agreement, if the communication graph $\mathcal{G}$ is connected, in at most D steps, where D is the diameter of the overlay network graph $\mathcal{G}$ [3]. Following this principle, in consensus-based decentralized auction algorithms, a conflict resolution strategy is used to converge on the assignment even when each vehicle is only partially informed on the status of other vehicles in the network (Figure 4).

In the next subsection we exploit the above introduced concepts to describe a policy-based decentralized solution to the spectrum allocation problem in the context of cognitive vehicular networks.

4.2 Max Consensus-based Auctions for Cognitive Spectrum Allocation

In this section we describe the Max-Consensus Auction (MCA) algorithm, applied to our spectrum allocation problem. MCA is a distributed algorithm that iterates between two (synchronous)[5] phases: a local bidding phase, where each vehicle greedily generates a bundle of frequencies, and a consensus phase, where conflict assignments are identified and resolved through local communication between neighboring vehicles. A diagram of the working principle of the MCA algorithm is depicted in Figure 4. In the rest of the chapter, we indicate with t an iteration of the algorithm, i.e., a single run of the two phases.

Procedure 3 MCA `LocalBidding` for vehicle i at iteration t

1: **Input:** $\mathbf{a}_i(t-1)$, $\mathbf{b}_i(t-1)$
2: **Output:** $\mathbf{a}_i(t)$, $\mathbf{b}_i(t)$, $\mathbf{m}_i(t)$
3: $\mathbf{a}_i(t) = \mathbf{a}_i(t-1)$, $\mathbf{b}_i(t) = \mathbf{b}_i(t-1)$, $\mathbf{m}_i(t) = \emptyset$
4: **if** `biddingIsNeeded`$(\mathbf{a}_i(t), \mathbb{T}_i)$ **then**
5:     **if** $\exists j : h_{ij} = \mathbb{I}(U_{ij}(t) > b_{ij}(t))$ **then**
6:        $j^\star = \underset{j \in \mathcal{J}}{\operatorname{argmax}} \{h_{ij} \cdot U_{ij}(t)\}$
7:        $\mathbf{m}_i(t) = \mathbf{m}_i(t) \oplus j^\star$
8:        $b_{ij^\star}(t) = U_{ij^\star}(t)$
9:        `update`$(j^\star, \mathbf{a}_i(t))$
10:    **end if**
11: **end if**

---

[5] In this chapter we focus on the simpler synchronous version, where the nth bidding phase cannot start if the $n^{th} - 1$ agreement phase is not completed. An example of asynchronous application of the MCA can be found in [8]. The main difference between synchronous and asynchronous version is the conflict resolution table, more complex if messages are allowed to arrive out of order.

Phase 1: Local Bidding

In Procedure 3 we describe the construction of the bundle of frequencies, the first phase of the MCA algorithm (Figure 4). In both phases (bidding and agreement or consensus), each vehicle $i \in \mathcal{I}$ keeps track of three vectors of states: an assignment vector $\mathbf{a}_i \in \{0 \cup \mathcal{J}\}^m$, which stores the identity of the vehicle assigned to each frequency in the network, a bid vector $\mathbf{b}_i \in \mathbb{R}^m$, which takes into account the maximum bid for each frequency, and a bundle vector $\mathbf{m}_i \in \{0 \cup \mathcal{J}\}^l$, which stores the frequencies assigned to vehicle i within its capacity l. These vectors are initialized at each iteration t (line 3 of Procedure 3). During the bidding phase, vehicles continuously sense frequencies that could potentially use, and if the policy allow it, they bid on multiple items to get permission to transmit on multiple available channels, until a stop condition is reached (line 4). The stopping condition is dictated by physical constraints, e.g., when a limit on the capacity does not allow to host (and so bid) on another frequency, or for policy constraints.

If the procedure allows a vehicle to bid on a given frequency, but such bid is too low to outbid the currently auction winner vehicle, the bidding phase terminates. If instead there is at least one biddable frequency, vehicle i seeks the best one to allocate according to its utility $U_{ij}(t)$. Each biddable frequency j that overbids its previous value b_{ij} (line 5) is stored in the temporary vector $h_{ij} = \mathbb{I}(U_{ij}(t) > b_{ij}(t))$, where $\mathbb{I}(\cdot)$ is an indicator function, unitary if the argument is true and 0 otherwise. Then, vehicle i appends *to the end* of its bid vector $\mathbf{m}_i$ the bid with the highest reward $j^\star$,

where:

$$j^\star = \underset{j \in \mathcal{J}}{\operatorname{argmax}} \{h_{ij} \cdot U_{ij}(t)\} \tag{22}$$

(lines 6–7) and updates the state vectors (lines 8–9). The order of frequencies obtained from subsequent bids gives a greedy nature to MCA and it is a crucial aspect to prove performance guarantees with respect to a Pareto optimal frequency allocation (see Section 4.4).

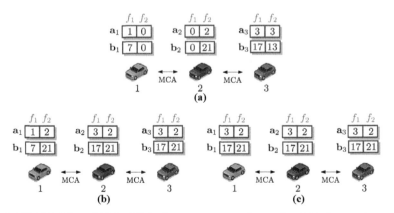

Figure 5. MCA Workflow: Vehicles bid for the temporary use of a set of available frequencies, assigning to each frequency a private utility value, assumed to be comparable across the vehicular network, for a distributed auction determination. In this example, three iterations of the MCA algorithm are enough to reach an agreement. The auction policy allows vehicles to bid on multiple frequencies per auction round.

Depending on the values of the configured policies, the functions `biddingIsNeeded()` and `update()` of Procedure 3 may behave differently. At the end of the first phase, the current bid vector $\mathbf{b}_i(t)$ and, if the auction policy allows it, the assignment vector $\mathbf{a}_i$ are exchanged with each neighbor $k \in \mathcal{N}_i$ of vehicle i.

Example 1 *(MCA Workflow). Consider Figure 5: three vehicles, with index set $\mathcal{I} = \{1, 2, 3\}$ detect two frequencies f_1 and f_2, with index set $\mathcal{J} = \{1, 2\}$ to be available. We assume a linear communication topology in which vehicle 1 communicates only with vehicle 2, vehicle 3 communicates only with vehicle 2, and vehicle 2 communicates with the others two vehicles.*

Assuming that all vehicles use as bids their residual budget, at iteration $t = 0$, vehicles 1, 2 and 3's initial bidding vectors are $\mathbf{b}_1(0) = (7, 0)$, $\mathbf{b}_2(0) = (0, 21)$, and $\mathbf{b}_3(0) = (17, 13)$. In this first bidding phase, vehicles assign themselves as winners for the frequency which they are bidding on, as they do not know yet any other bid; hence we have $\mathbf{a}_1(0) = (1, 0)$, $\mathbf{a}_2(0) = (0, 2)$, and $\mathbf{a}_3(0) = (3, 3)$ (see Figure 5a). After the first bid exchange, vehicle 1 updates both its bid and its assignment vector for the frequency 2, and vehicle 3 updates them for the frequency 2. Since vehicle 2 is now aware of all bids, it can update its bid and assignment vectors to the final values. Then, all vehicles move to the next iteration $t = 1$.

In iteration $t = 1$ the bidding phase does not change the bids values, i.e., $\mathbf{b}_1(1) = (7, 21)$, $\mathbf{b}_2(1) = (17, 21)$, and $\mathbf{b}_3(1) = (17, 21)$ and the agents exchange their states (Figure 5b). Each vehicle now is aware of all other bids.

In iteration $t = 2$, as no further bid changes are possible, the values of $\mathbf{b}_1(2) = (17, 21)$, $\mathbf{b}_2(2) = (17, 21)$, and $\mathbf{b}_3(2) = (17, 21)$ are confirmed, and a consensus on the final allocation is reached, that is, $\mathbf{a}_1(2) = (3, 2)$, $\mathbf{a}_2(2) = (3, 2)$, and $\mathbf{a}_3(2) = (3, 2)$ (see Figure 5c). Therefore, frequency 1 is allocated to vehicle 3 and frequency 2 to vehicle 2.

Phase 2 (Conflict Resolution)

After bids are independently made by each vehicle, they need to be exchanged to clear the auction and determine the winner(s). In the conflict resolution phase, vehicles use a maximum consensus strategy to converge to the winning bids $\bar{\mathbf{b}}$, and to compute the allocation vector $\bar{\mathbf{a}}$ (Procedure 4). The consensus strategy dictates that in a network, only the maximum bid is propagated along the network for each frequency. Intuitively, this means that after all vehicles have seen the highest bid only once, an agreement is reached. Formally, the consensus on the bid vector $\mathbf{b}_i$ after receiving the bids from each vehicle k in i's neighborhood $\mathcal{N}_i$ is obtained by comparing bid b_{ij} with b_{kj} for all k members of $\mathcal{N}_i$. This evaluation is performed by the function `IsUpdated()` (line 5).

In case the auction policy requires consensus only on a single frequency at a time, i.e., $|\mathbf{m}_i| = 1$, the function `IsUpdated()` merely checks if there is a higher bid, that is, if $\exists\, k, j : b_{kj} > b_{ij}$. This means that when a vehicle i receives from a neighboring vehicle k a higher bid for a frequency j, the receiver i always updates its bid vector $\mathbf{b}_i$ ($b_{ij} \leftarrow b_{kj}$), no matter when the higher bid was generated.

> **Remark 2** *Note that the MCA is a single shot auction, that is, once a vehicle has been outbid on a given frequency, it cannot rebid a second time, or a convergence on a conflict-free allocation may never be reached.*

When instead vehicles are allowed to bid on multiple frequencies in the same auction round (policy $|\mathbf{m}_i| > 1$), even if the received bid for a frequency is higher than what is currently known, the information received may not be up-to-date. In other words, the standard max-consensus strategy may not work. Each vehicle is required to evaluate the function `IsUpdated()`. In particular, `IsUpdated()` compares the time-stamps of the received bid vector and updates the bundle, the bid and the assignment vectors accordingly (Procedure 4, line 6). In general, i.e., when $|\mathbf{m}_i| > 1$, vehicles may receive higher bids which are out of date. When a vehicle receives a bid update, vehicle i has three options: (i) `ignore` the received bid leaving its bid vector and its allocation vector as they are, (ii) `update` according to the information received, i.e., $w_{ij} = w_{kj}$ and $b_{ij} = b_{kj}$, or (iii) `reset`, i.e., $w_{ij} = \emptyset$ and $b_{ij} = 0$. When $|\mathbf{m}_i| > 1$, the bids alone are not enough to determine the auction winner as frequencies can be released, and a vehicle i does not know if the bid received has been released or is outdated. The complete action set of the conflict resolution phase is shown in Table 1.

Table 1. Rules table for (synchronous) conflict resolution. The sender vehicle is denoted with k, and the receiver vehicle with i. The time vector **s** represents the time stamp of the last information update from each of the other vehicles.

$\mathbf{a}_{kj}$ for k is	$\mathbf{a}_{ij}$ for i is	Receiver's action (default leave)
k	i	if $b_{kj} > b_{ij} \rightarrow$ update
	k	update
	$m \notin \{i, k\}$	if $s_{km} > s_{im}$ or $b_{kj} > b_{ij} \rightarrow$ update
	none	update
i	i	leave
	k	reset
	$m \notin \{i, k\}$	if $s_{km} > s_{im} \rightarrow$ reset
	none	leave
$m \notin \{i, k\}$	i	if $s_{km} > s_{im}$ & $b_{kj} > b_{ij} \rightarrow$ update
	k	$s_{km} > s_{im} \rightarrow$ update else $\rightarrow$ reset
	$n \notin \{i, k, m\}$	if $s_{km} > s_{im}$ & $s_{kn} > s_{in} \rightarrow$ update if $s_{km} > s_{im}$ & $b_{kj} > b_{ij} \rightarrow$ update if $s_{kn} > s_{in}$ & $s_{im} > s_{km} \rightarrow$ reset
	none	if $s_{km} > s_{im} \rightarrow$ update
none	i	leave
	k	update
	$m \notin \{i, k\}$	if $s_{km} > s_{im} \rightarrow$ update
	none	leave

4.3 Max-Consensus Auction Policies

In the previous section we have seen how the MCA mechanism works, and how it is used by vehicles to reach an agreement that maximizes the network utility. Such utility may differ depending on the context and the application, i.e., it is one of the variant aspects of the MCA mechanism. In this section we give few examples of policies that can be instantiated to modify the behavior of the spectrum allocation algorithm in our cognitive vehicular network.

Utility Function Policy: the most intuitive example of policy is the private utility function that vehicles use to bid on the available spectrum. Delay-sensitive cognitive vehicular network applications such as online-gaming may have a different utility than a bandwidth-sensitive application such as a peer-to-peer based vehicle security firmware upgrade, in which it is more important to minimize the content distribution time of the upgrade across a fleet of vehicles. Furthermore, in energy saving applications, it may be more important to avoid battery waste from activating multiple radio channels simultaneously.

Procedure 4 MCA ConflictResolution for vehicle i at iteration t

---

1: **Input:** $a_i(t)$, $b_i(t)$, $m_i(t)$
2: **Output:** $a_i(t)$, $b_i(t)$, $m_i(t)$
3: **for all** $k \in \mathcal{N}_i$ **do**
4:   **for all** $j \in \mathcal{J}$ **do**
5:     **if** IsUpdated(b_{kj}) **then**
6:       update($b_i(t)$, $a_i(t)$, $m_i(t)$)
7:     **end if**
8:   **end for**
9: **end for**

Auction Policies: the MCA policies are used not only to handle different allocation problems, but also to obtain different algorithmic behaviors. Let us consider the number of simultaneously frequencies (i.e., channels) allowed to be assigned in a single auction round (auction policy). Classical mechanism design studies have shown how we can enforce a truthful auction: in particular, it is known how single item auctions have the property of being *strategy-proof*, if combined with a second price auction [22]. That is, vehicles may have incentives to bid on frequencies using their truthful evaluation, if they do not have incentives to save for future higher bids (see the following Remark).

Remark 3 *Although second price auctions are algorithmically interesting, in our settings an auction on all available frequencies may not be applicable for two reasons: first, we assumed that vehicles cooperate to reach a Pareto optimal global utility (and hence do not compete), and second, because the spectrum is likely to be freed up (and so available for bid again) soon after an agreement phase, bidders may preserve resources for stronger future bids. The MCA mechanism presented can however inspire interesting research directions; for example, vehicles could*

> *act as selfish users, and collect payments of some sort to opportunistically use a given frequency to allow or deny communications on given flows. We leave the investigation of pricing mechanisms for spectrum allocation in cognitive network protocols as an interesting open question.*

Privacy Mode Policy: another example of policy is the *privacy mode* of the vector $\mathbf{a}_i$, that is, a vector that keeps track of the identities of the current vehicle hosting a given frequency during an epoque; $\mathbf{a}_i$ may assume two forms: *least* and *most informative*.

In its least informative form, $\mathbf{a}_i \equiv \mathbf{x}_i$ is a binary vector where x_{ij} is equal to one if vehicle i hosts frequency j and 0 otherwise. In its most informative form, $\mathbf{a}_i \equiv \mathbf{w}_i$ is a vector of vehicles IDs that are so far winning the spectrum; w_{ij} represents the identifier of the vehicle that made the highest bid so far to host frequency j. Note that when $\mathbf{a}_i \equiv \mathbf{w}_i$ the assignment vector reveals information on which vehicles are so far the winners of the "auction", whereas if $\mathbf{a}_i \equiv \mathbf{x}_i$ vehicle i only knows if it is winning each frequency or not. As a direct consequence of the max-consensus, this implies that when the privacy policy is set to its least informative form, each vehicle only knows the value of the maximum bid so far without knowing the identity of the bidder. Setting the privacy policy to its least informative form requires a service provider to orchestrate the actual frequency allocation; the decentralized service provider would be the only entity with knowledge of the vehicles that won the auction, while a privacy policy set to its most informative form enables a fully distributed allocation.

4.4 Max-Consensus Auction Properties

In this section we summarize some theoretical results on the MCA algorithm. First we analyze the convergence of the algorithm, i.e., a conflict-free assignment is found in a finite number of steps, and later, leveraging well-known results on sub-modular functions [25, 12], we show that under the assumption of pseudo sub-modularity of the utility function, MCA guarantees a $(1 - \frac{1}{e})$ optimal approximation, that is, a better approximation does not exist unless $P = NP$.

Pseudo Sub-modularity

The MCA algorithm is proven to exhibit some convergence and performance properties, under the assumption that the utility function U_{ij} appears to be sub-modular to other bidders [17]. Sub-modularity is a well studied concept in mathematics [25], and it can be defined as follows:

Definition 1 *(sub-modular function). The marginal utility function $U(j, \mathbf{m})$ obtained by adding a object j to an existing bundle $\mathbf{m}$, is sub-modular if and only if*

$$U(j, \mathbf{m}') \geq U(j, \mathbf{m}) \quad \forall \mathbf{m}' | \mathbf{m}' \subset \mathbf{m}. \tag{23}$$

This means that if a vehicle uses a sub-modular utility function, a value of a particular frequency j cannot increase because of the presence of other frequencies in the bundle.

The notion of sub-modularity is a key aspect of MCA: releasing outdated bids without a sub-modular utility may break the convergence to a conflict-free assignment [5, 8, 9]. Although having sub-modular utility functions may be realistic in many resource allocation problems [20], in the distributed spectrum allocation problem this assumption may be too restrictive, as the value of a frequency may increase as new frequencies are added to the bundle. To guarantee convergence without using a sub-modular utility function, as in [8, 17], we let each vehicle communicate its bid on frequency j obtained from a bid warping function:

$$\mathcal{W}_{ij}(U_{ij}, \mathbf{b}_i) = \min_{k \in \{1,\ldots,|\mathbf{b}_i|\}} \{U_{ij}, \overline{\mathcal{W}}_{ik}\} \tag{24}$$

where $\overline{\mathcal{W}}_{ik}$ is the value of the warping function for the kth element of $\mathbf{b}i$. Note how by definition, applying the function $\mathcal{W}$ to the bid before sending it is equivalent to communicating a bid that is never higher than any previously communicated bids. In other words, bids *appear* to other vehicles to be obtained from a sub-modular utility function.

4.5 Convergence to a Conflict-free Spectrum Allocation

During the communication between the two phases of MCA, vehicles exchange their bids with only their first-hop neighbors, therefore a change of bid information needs to necessarily traverse all the communication network, which we assume has diameter D. The following proposition (Proposition 4.1) states that a propagation time of D hops is also a sufficient condition to reach max-consensus on a single frequency allocation. Another interesting observation which follows from the result is that, given m frequencies to allocate, the number of steps for MCA to converge is always $D \cdot m$ in the worst case, regardless of the size of the bundle vector. This means that the same worst-case convergence bound is achieved if MCA runs on a single or on multiple frequencies simultaneously. Therefore, by induction on the size of the bundle the following result holds (as a corollary of Theorem 1 in Choi et al. [5]):

Proposition 4.1 *(Convergence of MCA). Given a set of m frequencies to be allocated to n vehicles that communicate over a network of diameter D, the utility function of each vehicle is pseudo sub-modular, and the communications occur over reliable channels, then the MCA converges in a number of iterations bounded above by D · m.*

Proof 4.1 *We use $\mathcal{W}_{ij}(U_{ij}, \mathbf{b}_i)$ as a bid function (sub-modular by definition). From [5] we know that a consensus-based auction run by a fleet of N_u agents, each assigned at most L_t objects, so as to allocate N_t objects, converges in at most $N_{min} \cdot D$ where $N_{min} = \min\{N_t, N_u \cdot L_t\}$. Note that the proof of Theorem 1 in [5] is independent of the utility function used by the agents as long as they are sub-modular, and of the constraints that need to be enforced on the objects. Since for MCA to converge, every frequency needs to be assigned, in the distributed spectrum allocation problem, N_{min} is always equal to $N_t \equiv m$, and therefore we prove the claim.[6]*

---

[6] Note that his proof is equivalent to the one shown for the virtual network embedding auctions at [8].

As a direct corollary of Proposition 4.1, we compute a bound on the number of messages that vehicles have to exchange in order to reach an agreement on the spectrum allocation. Because we only need to traverse the communication network once, the following result holds:

Corollary 4.1 *(Communication overhead). The number of messages exchanged to reach an agreement on the spectrum allocation using the MCA mechanism is at most $D \cdot \varepsilon \cdot m$, where D is the diameter of the physical network, ε is the number of directed edges in the vehicular network, and m is the number of frequencies to allocate.*

4.6 Performance Guarantees

As shown in the previous section, distributed auctions converge to a solution if the bidding function has a property of sub-modularity. Thus, assuming that our bids on frequencies are obtained using a pseudo sub-modular function, we obtain the following result on MCA performance guarantee.

Theorem 4.1 *(MCA Approximation). The MCA mechanism yields an $(1 - \frac{1}{e})$-approximation with respect to the optimal spectrum allocation solution.*

Proof 4.2 *The MCA consensus approach assumes that each vehicle i does not bid on a frequency j unless it brings a positive utility, therefore U_{ij} and so W_{ij} are positive. Moreover, if we append to the bid vector $\mathbf{b}_i$ an additional set of frequencies $\mathbf{v}$ resulting in bid vector $\mathbf{b}'_i$, we have:*

$$W_{ij}(U_{ij}, \mathbf{b}'_i) \leq W_{ij}(U_{ij}, \mathbf{b}_i) \ \forall \ \mathbf{v} \neq \emptyset \tag{25}$$

which means that W_{ij} is monotonically non-increasing. Since the sum of the utilities of each vehicle, and since the bid warping function $W_{ij}(U_{ij}, \mathbf{b}_i)$ of MCA is a positive, monotone (non-increasing) and sub-modular function, all the axioms of Theorem 1 in Nemhauser et al. [25] on sub-modular functions are satisfied. Therefore the claim holds.

5. Conclusions

A cognitive vehicular network may have not only variable channel availability, but variable contact rate, variable hop-to-top and end-to-end delays, variable vehicle traffic density, and may be required to run applications with different or conflicting goals. In such complex heterogeneous and dynamic scenarios, a "one-size-fits-all" spectrum allocation solution possibly cannot exist. To this end, after modeling the spectrum allocation problem with optimization theory as a network utility maximization, we described two alternative and complementary policy-based approaches that allow cognitive radio-equipped vehicles to tune the "knobs" of the spectrum allocation mechanisms.

In particular, in the first part of the chapter we described a technique that leverages primal and dual decomposition, or a combination of the two, to adapt the multi-dimensional spectrum allocation problem to a global service goal or objective. In the second part of the chapter, we used Max-Consensus Auctions to describe another policy-based distributed spectrum allocation. Under reasonable assumptions, we have shown how the latter approach provides guarantees on both convergence time and performance, with respect to a Pareto optimal spectrum allocation.

References

[1] Ian F. Akyildiz, Won-Yeol Lee, Mehmet C. Vuran and Shantidev Mohanty. Next generation/dynamic spectrum access/cognitive radio wireless networks: A survey. Computer Networks Journal (Elsevier), 50: 2127–2159, 2006.

[2] D.P. Bertsekas. The auction algorithm: A distributed relaxation method for the assignment problem. Annals of Operations Research, 14(1): 105–123, 1988.

[3] A. Bondy and U.S.R. Murty. Graph Theory. Graduate Texts in Mathematics. Springer, 2008.

[4] S. Boyd and L. Vandenberghe. Convex Optimization. http://www.stanford.edu/people/boyd/cvxbook. html, 2004.

[5] Han-Lim Choi, L. Brunet and J.P. How. Consensus-based decentralized auctions for robust task allocation. IEEE Transactions on Robotics, 25(4): 912–926, Aug 2009.

[6] Thomas H. Cormen, Charles E. Leiserson, Ronald L. Rivest and Clifford Stein. Introduction to Algorithms (2nd ed.). MIT Press and McGrawHill, 2001.

[7] K. Cumanan, R. Krishna, Z. Xiong and S. Lambotharan. Multiuser spatial multiplexing techniques with constraints on interference temperature for cognitive radio networks. IET Signal Processing, 4(6): 666–672, 2010.

[8] Flavio Esposito, Donato Di Paola and Ibrahim Matta. On distributed virtual network embedding with guarantees. ACM/IEEE Transactions on Networking. Nov 2014, http://dx.doi.org/10.1109/ TNET.2014.2375826.

[9] Flavio Esposito, Donato Di Paola and Ibrahim Matta. A General Distributed Approach to Slice Embedding with Guarantees. In Proc. of the IFIP Networking, Brooklyn, NY, USA, 2013.

[10] Flavio Esposito, Anna Maria Vegni, Ibrahim Matta and Andrea Neri. On modeling speed-based vertical handovers in vehicular networks. In GLOBECOM Workshops (GC Wkshps), 2010 IEEE, pp. 11–15, Dec 2010.

[11] S. Even, A. Itai and A. Shamir. On the Complexity of Timetable and Multicommodity Flow Problems. SIAM Journal on Computing, 4(5): 691–703, 1976.

[12] Uriel Feige. A Threshold of ln n for Approximating Set Cover. J. ACM, 45(4): 634–652, July 1998.

[13] B.P. Gerkey and M.J. Mataric. Sold!: auction methods for multirobot coordination. IEEE Transactions on Robotics and Automation, 18(5): 758–768, Oct 2002.

[14] S. Giannini, D. Di Paola, A. Petitti and A. Rizzo. On the convergence of the max-consensus protocol with asynchronous updates. In IEEE 52nd Annual Conference on Decision and Control (CDC), 2013, pp. 2605–2610, Dec 2013.

[15] Simon Haykin. Cognitive radio: brain-empowered wireless communications. IEEE Journal on Selected Areas in Communications, 23(2): 201–220, Feb 2005.

[16] Dorit S. Hochbaum. Approximation algorithms for np-hard problems. Chapter Approximating Covering and Packing Problems: set cover, vertex cover, independent set, and related problems, pp. 94–143. PWS Publishing Co., Boston, MA, USA, 1997.

[17] Luke B. Johnson, Han-Lim Choi, Sameera S. Ponda and Jonathan P. How. Allowing Non-Submodular Score Functions in Distributed Task Allocation. In IEEE Conference on Decision and Control (CDC), 2012.

[18] Xin Kang, Ying-Chang Liang, Hari Krishna Garg and Lan Zhang. Sensing-based spectrum sharing in cognitive radio networks. In GLOBECOM, pp. 4401–4405. IEEE, 2008.

[19] George Karakostas. Faster approximation schemes for fractional multicommodity flow problems. In Proceedings of the thirteenth annual ACM-SIAM symposium on Discrete algorithms, page 166173. SIAM Press, 2002.

[20] Ariel Kulik, Hadas Shachnai and Tami Tamir. Maximizing submodular set functions subject to multiple linear constraints. In Proceedings of the Twentieth Annual ACM-SIAM Symposium on Discrete Algorithms, SODA '09, pp. 545–554, Philadelphia, PA, USA, 2009. Society for Industrial and Applied Mathematics.

[21] T. Lemaire, R. Alami and S. Lacroix. A distributed tasks allocation scheme in multi-uav context. In IEEE International Conference on Robotics and Automation, 2004. Proceedings. ICRA '04. 2004, 4: 3622–3627 Vol. 4, April 2004.

[22] Brendan Lucier, Renato Paes Leme and Eva Tardos. On revenue in the generalized second price auction. In Proceedings of the 21st International Conference on World Wide Web, WWW '12, pp. 361–370, New York, NY, USA, 2012. ACM.

[23] Nancy A. Lynch. Distributed Algorithms. Morgan Kaufmann, 1st edition, March 1996.

[24] Saber Mirzaei and Flavio Esposito. An alloy verification model for consensus-based auction protocols. CoRR, abs/1407.5074, 2014.

[25] G. Nemhauser, L. Wolsey and M. Fisher. An analysis of approximations for maximizing submodular Set Functions. Math. Prog. 14(1): 265–294, 1978.

[26] M. Nguyen and H. Lee. Effective scheduling in infrastructure-based cognitive radio networks. IEEE Transactions of Mobile Computing, 10(1): 853867, 2011.

[27] A. Petitti, D. Di Paola, A. Rizzo and G. Cicirelli. Consensus-based distributed estimation for target tracking in heterogeneous sensor networks. In Decision and Control and European Control Conference (CDC-ECC), 2011 50th IEEE Conference on, pp. 6648–6653, Dec 2011.

[28] Wei Ren and Randal W. Beard. Distributed Consensus in Multi-vehicle Cooperative Control: Theory and Applications (Communications and Control Engineering). Springer, 1 edition, December 2007.

[29] Wei Ren and Randal W. Beard. Distributed Consensus in Multi-vehicle Cooperative Control: Theory and Applications. Communications and Control Engineering. Springer, London, 2008.

[30] S.L. Smith and F. Bullo. Monotonic target assignment for robotic networks. Automatic Control, IEEE Transactions on, 54(9): 2042–2057, Sept 2009.

[31] Ye Wang, Qinyu Zhang, Yalin Zhang and Peipei Chen. Adaptive resource allocation for cognitive radio networks with multiple primary networks. EURASIP Journal on Wireless Communications and Networking, 2012(1), 2012.

[32] M.M. Zavlanos, L. Spesivtsev and G.J. Pappas. A distributed auction algorithm for the assignment problem. In 47th IEEE Conference on Decision and Control, CDC, pp. 1212–1217, Dec 2008.

4

Data Dissemination in Cognitive Vehicular Networks

Andre L.L. Aquino,* Rhudney Simões, David H.S. Lima and
Heitor S. Ramos

ABSTRACT

Cognitive radio technology is an efficient solution to enhance the spectral utilization, which takes advantage of residual resources of underutilized channels by an opportunistic use of the spectrum. A cognitive vehicular network allows the cognitive radio to benefit from holes or white spaces presented in the spectrum and admits an opportunistic access that searches for the best wireless communication channel available. This Chapter presents an overview of important technologies used to allow data dissemination in these networks. The topics herein addressed are: (i) data dissemination and communication; (ii) general broadcasting protocols; and (iii) potential data dissemination applications.

1. Introduction

A Vehicular Ad Hoc Network (VANET) is a network where each node represents a vehicle equipped with wireless communication technology [1, 15]. The communication in these networks generally follows two different models: Vehicle-to-Vehicle (V2V), when vehicles communicate directly, and Vehicle-to-Infrastructure (V2I), when vehicles exchange information with roadside units such as access points or any other network infrastructure. Some general scenarios where VANETs can emerge are illustrated in Figure 1 [28]. The first scenario represents areas with low node density, such as highways, where communication employs opportunistic forwarding,

---
SensorNet-UFAL Research Group, Computer Institute, Federal University of Alagoas – Brazil.
   Emails: rhudney.simoes@gmail.com; dhs.lima@gmail.com; heitor@ic.ufal.br
* Corresponding author: alla.lins@pq.cnpq.br; alla@ic.ufal.br

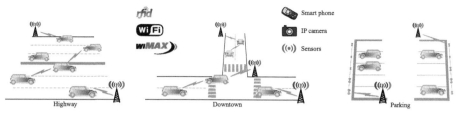

Figure 1. Intelligent vehicular sensor system [28].

i.e., information is transmitted when two nodes are within each others transmission range. The second and third scenarios illustrate urban areas, where communication may occur using a mix of communication technologies.

In scenarios such as the ones showed in Figure 1, data dissemination in VANETs is a crucial aspect [8]. The concept of data dissemination in VANETs refers to the spreading of information through broadcast over a wireless distributed network [33]. A reliable data dissemination does not depend only on the broadcast protocol. An important task is the choice of an useful channel.

To allow an efficient data dissemination in VANETs, it is supposed to employ a fixed spectrum band allocation such as the band assigned to the Dedicated Short-Range Communication (DSRC) standard (75 MHz spectrum allocated in 5.9 GHz band), unlicensed band (for instance 2.4 GHz), industry, science, and medicine (ISM) band, or a wide-band using overlay techniques like ultrawideband (UWB) signaling [39]. In this way, Cognitive Radio (CR) [16] has emerged as a key technology, which enables VANETs to access both licensed and license-exempt spectrum bands in an opportunistic manner.

The CR technology is an efficient solution to enhance the spectral utilization, which takes advantage of residual resources of underutilized channels by an opportunistic use of the spectrum, so it comes to solve problems of overcrowding and bandwidth scarcity. A Cognitive Vehicular Network (CVN), allows the CRs allows the cognitive radio to benefit from holes or white spaces presented in the spectrum and admits an opportunistic access that searches for the best wireless communication channel available [18]. Vehicles equipped with CR need to uninterruptedly monitor the channel in order to verify presence of primary users signals. There is no guarantee that vehicles will be able to complete their transmissions because the original primary users signals may arrive and, occasionally, interrupt an current transmission. In this such case, the CR vehicles need to vacate the channel.

As discussed by Vegni et al. [51], vehicles in a CVN have many unique characteristics that involve additional considerations than merely placing a CR within a vehicle. For instance, the spectrum availability perceived by each moving vehicle changes dynamically over time, as a function not only of the activities of the licensed or primary users but also based on the relative motion among them. Spectrum measurements need to be undertaken over the general movement path of the vehicles, leading to a path-specific distribution, instead of focusing only on the temporal dimension.

Additionally, Vegni et al. argue that vehicles in a CVN can also leverage the constrained nature of motion, i.e., vehicles are restricted to linear and predictable paths corresponding to streets and freeways. At busy hours or in urban areas, spectrum information can be exchanged over multiple collaborative vehicles, improving the knowledge about the spectrum availability.

The CVNs could be classified into three different classes differentiated by their communication elements [13, 51]: (i) V2V communication occurs when the heterogeneous features are complex to keep a stable link, so the vehicles have to communicate among them. The dynamic topology results in frequent disconnections; (ii) V2I communication occurs when the vehicles communicate to roadside infrastructure or other vehicles to forward the data to be disseminated; and (iii) centralized communication occurs when the vehicle communicate only to the roadside unit, i.e., the data dissemination must be performed through the centralized infrastructure.

The remain of this chapter presents an overview about important technologies used to allow the data dissemination in CVNs. The subjects herein addressed are: in Section 2, we present important issues about data dissemination and communication. In Section 3 we discuss about broadcast protocols to improve data dissemination. In Section 4 we present some neighbors discovery algorithms for cognitive radio. In Section 5 we show some applications where data dissemination features an important role and the CR can be satisfactorily applied. Finally, in Section 6 we present the final remarks.

2. Data Dissemination and Communication Issues

The data dissemination problem in wireless environments was initially studied in infrastructured environments to guarantee a good performance in general mobile applications. As presented by Rao and Reddy [40], data dissemination in these environments is performed over asymmetrical communication, where the downlink capacity is much greater than the uplink one. In this architecture there is a stationary server continuously broadcasting different data items over the air. Mobile clients continuously listen to the channel and access the data of their interest whenever it appears on the channel. An important issue in this type of data dissemination is how fast mobile clients access the data item of their interest, i.e., the minimum access time so that mobile clients can save battery. Rao and Reddy discuss different methods of data dissemination in infrastructured environments, distinguished as follow:

- *Push-based data dissemination*: In this case, the server broadcasts data proactively to all clients according to previous responses of a scheduling algorithm. The broadcast strategy determines the order and frequencies of the propagated data.

- *On-demand data dissemination*: In this case, also called pull-based data dissemination, when a client needs a data item, it sends to the server an on demand request for the item through its uplink. Client requests are queued up (if necessary) at the server upon arrival. The server repeatedly chooses an item from among the requests, broadcasts it over the broadcast channel, and removes the associated request from the queue. The clients monitor the broadcast channel and retrieve the required items.

- *Data allocation over multiple broadcast channels*: In this case, it is used a data-scheduling algorithm, which works dynamically, and allocates data according to changing access patterns to achieve efficient data access and channel utilization so that the access time is minimum.

Nowadays, with the advent of new mobile applications based on ad hoc communication, such as: cellphones in ad hoc mode, wireless sensor, or vehicular applications, the infrastructure-based communication paradigm are not longer enough for data dissemination scenarios. Some data dissemination techniques are proposed to general mobile ad hoc application. However, each application has particular requirements and consequently we observe a complexity increasing of data dissemination protocols.

In particular, as presented by Dubey et al. [12], VANETs data transfer is performed in a multi-hop communication in which high speed vehicles are acting as the data carrier. Vehicles are constrained to move on definite path restricted to the road layout and the traffic conditions. In VANETs, multi-hop data delivery is a complex task due to the high mobility and frequent disconnections occurring in the vehicular networks. A big challenge in VANETs scenario is the collection of information about the road or traffic conditions for the safety and convenience purpose. In many dissemination techniques, the vehicle carries the packet until it finds any other vehicle in his range which is moving towards the direction of the destination, and then, it forwards the packet to that vehicle. Since roads layout are already defined, the vehicle selects the next road having minimum latency to forward the packet to the destination. However, the main goal is to estimate which path should be followed to minimize the delay, so that the bandwidth can be efficiently used. Some authors [8, 12, 24] discuss different methods of data dissemination in VANETs distinguished as follow:

- *Push-based data dissemination*: It is similar to infrastructure one where the data is managed by a data center which collects the data from the outside world and make it ready to deliver to the vehicles. An example of this protocol is presented by Zao et al. [56]. They propose a data pouring and buffering scheme for push-based data dissemination. This scheme selects one or some road having high density and mobility of vehicles. Relay and broadcast stations are placed at the intersection points and used to store data at the intersections. In this scheme the data has been transferred from data center to the buffers present at the intersections by this way the availability of the data is increased at the intersection, the load on the server is reduced and data delivery ratio is increased. There may be possibility of collision between the new data item send by data center and broadcast data by relays nodes. To avoid this collision, broadcast period is divided into two slots.

- *On-demand data dissemination*: Also similar to infrastructure one where it is used by vehicles to query the data for the specific response from data center or from other vehicles. In this scheme the data is managed by both the data center and vehicles which are moving on the road. When a vehicle needs any data query it firstly sends a beacon message to find the list of neighbors. These vehicles could be equipped with digital maps, having street level maps and traffic details like traffic density and vehicle speed on roads at different times. This approach uses a store-and-forward mechanism to deliver the data [29]. In this mechanism data packets are carried by vehicles and it forward that packet whenever they meet other vehicle moving towards destination. An example of this protocol is presented by Zao and Cao [55]. They proposed the Vehicular Assisted Data Dissemination (VADD) scheme for pull-based data dissemination. In this protocol when the data has to be forwarded from one place to another place then it suggests that path selection should be done on the basis of high density of vehicle.

• *Hybrid data dissemination*: Along with the push and pull models that were presented, there are few schemes that combine both models in order to support different types of applications within a VANET environment. An example of this protocol is presented by Dikaiakos and Nadeem [11]. The proposed protocol supports the establishment of distributed service infrastructure over VANETs, by specifying the syntax and the semantic of messages between vehicles. It uses methods from both dissemination models: a push-based technique is used for safety messages such as alerts about emergencies or hazardous traffic conditions; and a pull-based technique is proposed to retrieve information by location-sensitive queries issued by vehicles on demand.

CVNs are a particular case of VANETs that considers the usage of CRs to perform the communication. The vehicles could communicate through licensed and license-exempt spectrum bands. Therefore, in the context of CVNs, reliable data dissemination is much more challenging than traditional wireless networks. As mentioned by Singh et al. [48], the effect of vehicular mobility on spectrum management show that the primary users may have to face adverse interference due to the incorrect detection of occupied frequencies as a result of the Doppler spread generated by vehicular mobility. At the same time, a CR-enabled vehicle can gather samples of signal at different locations along its path while moving. This feature increases the spatio-temporal diversity of the samples and, in turn, decreases the chances of inaccurate decisions due to shadowing effects. In additional, Singh et al. comment that V2V communication is more complex and difficult because of dynamically evolving topologies and frequent disconnections that are due to vehicles going out of range. This happens more often when the vehicular density is low. Such disconnections can create instability and performance issues for upper layer protocols such as routing and transport. For routing in vehicular networks, some approaches based on Delay Tolerant Networking (DTN) are becoming popular [10]. The main idea of theses networks is to perform an opportunistic communication when vehicles are in contact.

In addition to the already known issues of wireless environments, important aspects in CVNs must be considered: the diversity in the number of channels each cognitive node can use adds another challenge by limiting node's accessibility to its neighbors; CR nodes have to compete with the Primary Radio (PR) nodes for the residual resources on many channels and opportunistically use them; during communication, CR nodes should communicate in such a way that it should not degrade the reception quality of PR nodes by causing CR-to-PR interference; and CR nodes should immediately interrupt its transmission whenever a neighboring PR activity is detected [43]. In addition, the data transfer in CVNs occurs with vehicles moving in high speed, so the coordination among vehicles is hard to achieve. No central entity is used to regulate the access over channels, thus, reliable data dissemination is even more complex.

The strategies used to disseminate data in CVNs are similar to the ones used in VANETs and, consequently, present similar issues. However, the previously mentioned protocols to VANETs have to be adapted to achieve some CRs requirements. Some aspects that must be considered are: the V2V communication, and the mobility of CR nodes. These issues are not common in traditional Cognitive Radio Networks (CRNs). These differences are directly related to the communication interface used to allow a robust data dissemination.

In this way, there are some efforts to allow the use of license-exempt spectrum bands by a communication interface in 4G/5G technologies. Hussain and Sharawi [20] comment about an initial proposal to use Multiple–input-multiple–output (MIMO) technology in 4G wireless standards to meet high data rate requirements in multipath wireless channels. Hussain and Sharawi show a compact meandered-line planar reconfigurable MIMO antenna for CRs applications. This two-element MIMO system was made tunable in different bands using PIN and varactor diodes. The most distinguishing feature of the presented design is its planar structure with operation at lower frequency bands starting from 580 ~ 680 and 834 ~ 1120 MHz by using varactor diode tuning. Other works also present MIMO frequency-reconfigurable antennas without the sensing antenna for CR applications [19, 25].

Zhang et al. [54] discuss that emerging 5G wireless networks aims at ensuring that various contemporary wireless applications can be timely and satisfactorily served at any time and any place, and in any way. One of the most important services in 5G wireless applications is the bandwidth-intensive and time-sensitive multimedia, even including 3D immersive media. Such transmissions were previously confined to wired networks such as the Internet, but are now making forays into mobile devices and wireless networks. Zhang et al. argue that to support these highly bandwidth-intensive and time-sensitive multimedia services for the emerging 5G wireless networks, the usage of CRs could be a promised technology among other ones. For instance, hybrid CRNs have been proposed for adoption in cellular networks to explore additional bands and expand the capacity. We agree that the adoption of CRs in 5G wireless networks could improve some important issues in CVNs, as previously mentioned.

While these technologies are not in use in all environments, to allow the data dissemination in CVNs, some works improve the VANETs data dissemination approaches by considering some modifications in the communication interface, when the CRs is used. For instance, some works propose to modify the broadcasting strategies, neighbors discovery or channel selection strategies. Specifically to channel selection strategies, the SURF protocol [43], initially proposed to CRNs but used in CVNs, focus on the selection of the best available channel allowing a better data dissemination process. Its goal is to ensure reliable contention-aware data dissemination. Usually channel selection strategies provide a way to nodes to select channels for transmission. On the contrary, CR nodes select the best channel not only for transmission but also for overhearing. As a result, both sender and receiver are tuned to the most suitable channel for effective and reliable data dissemination.

An important issue regarding data dissemination in VANETs, and consequently in CVNs, is the broadcast storm [45]. Broadcast is widely used in all kind of networks, many protocols use it to disseminate control and data packets. The broadcast storm problem must be treated to avoid a big number of redundant messages. It happens when various vehicles rebroadcast the same packet causing packet collisions. Therefore, broadcast needs to be used carefully. For instance, algorithms based on flooding is not appropriate to dense areas.

Specifically, in CVNs the broadcast storm may happen in many other situations, for instance, when broadcasting control messages for neighbor discovery, and when broadcasting packets from disconnected available channels. As presented by Rehmani [41], in CVNs, the broadcasting is performed more frequently than in multi-channel wireless networks, where the availability of multiple channels are static. On the contrary, channels are dynamic in CVNs due to the primary radio activity. Therefore, the broadcast storm problem is present in these networks. The strategies used in general ad hoc networks to avoid this problem and still suitable for CVNs, are: (i) to reduce the possibility of redundant rebroadcast by selecting groups of relevant nodes to be forwarder candidates; and (ii) to differentiate the timing of rebroadcast by using a duty cycle strategy [36].

Finally, another important aspects regarding CRs, and consequently in CVNs, are security issues [45]. In this case general strategies used in CRs can be directly applied for CVNs. Two common attacks that can occur in data dissemination application of CVNs are:

- *Denial-of-service*: This attack makes the resources of a system unavailable for the users. In this case, unlicensed users pretend to be primary user to consume the channel in a selfish way, this attack is also called *Primary User Emulation Attack* (PUEA). A method to prevent this attack is differentiating the original primary user signal to the fake signal emitted by secondary one. Chen and Park [9] show a position based method used to identify the attacker, and Mathur and Subbalakshmi [31] use public key encryption based to identify the primary user.

- *Jammers*: This attack compromises the use of the channel, causing packet delay and disconnection. Jamming signals are used to cause radio interference. Jammers may be classified into several types, and the most aggressive is the reactive one. The reactive jammer remains quiet while there is no legitimate signal activity on the channel, after realizes the presence of the signal, the jammer immediately starts to transmit a random signal to cause channel disturbance. The consequences of this attack are the loss of link reliability, increased energy consumption, extended packet delay, and disruption of end routes [46].

3. Broadcasting in CVN

Data dissemination in CVNs is destined for public interest, and not only for an individual. Therefore, a broadcasting scheme is more suitable when compared to routing approach that employs an unicasting process [27]. Broadcasting scheme has the advantage that a vehicle does not require a destination address neither a route to a particular destination. This aspect avoids complex mechanisms to route discovery, address resolution, and topology management [24].

Broadcast is a process that disseminates information from a source to all receivers subscribed to a service. It is a fundamental operation widely applied for many communications systems, furthermore, it is the base of data dissemination strategies. In general wireless networks, the broadcast protocols are commonly classified into stateful or stateless [42] as depicted in Figure 2 and explained below.

- *Stateful:* broadcasting algorithms that require that nodes have the knowledge of the network topology in their local neighborhood. This is commonly achieved by proactively exchanging messages between neighbors.
  - ○ *Neighbor designated:* A node that transmits a packet specifies which neighbor should forward the packet.
  - ○ *Self-pruning:* A node receiving a packet will decide itself if the packet will be forwarded.
  - ○ *Energy-efficient:* It is considered the residual energy to forward the packet.
- *Stateless:* broadcasting algorithms that do not require any knowledge of the neighborhood.
  - ○ *Probability-based:* Each node forwards the packet considering a probability p and drops the packet with a probability $1 - p$.
  - ○ *Simple flooding:* The packet is always forwarded.
  - ○ *Location-based:* The node forwards the packet considering a distance between itself and the receiver.

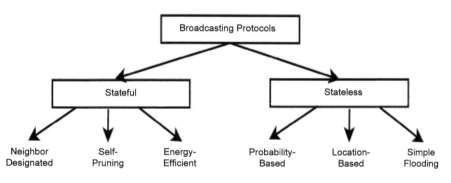

Figure 2. Broadcasting protocol classification [42].

Based on this classification, general broadcasting approaches to CVNs are stateless. Basically because the complete or partial knowledge of the network topology can be prohibitive. To allow the data dissemination in CVNs, broadcasting strategies have to be adapted to achieve some CRs requirements. In this way, broadcasting strategies applied to CRNs could be applied directly in CVNs. The aspects that must be adequate are: V2V communication, and the mobility of CR nodes.

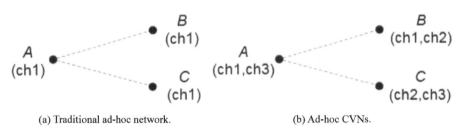

(a) Traditional ad-hoc network.   (b) Ad-hoc CVNs.

Figure 3. Single-channel and multi-channel broadcast problems [49].

Therefore, as explained by Song and Xie [49], unlike traditional ad hoc networks where the channel availability is uniform (single-channel or multi-channel), in CRNs, different secondary users may acquire different sets of available channels. This non-uniform channel availability imposes special design challenges for broadcasting strategies. Song and Xie report that for traditional single-channel and multi-channel ad hoc networks all nodes can tune to the same channel. Thus, broadcast messages can be conveyed through a single common channel which can be heard by all nodes in a network. However, in CRNs, the availability of a common channel for all nodes may not exist. Therefore, broadcasting messages on a global common channel is not feasible in CRNs.

The above problem is illustrated by Song and Xie in Figure 3, where it is presented the challenges of broadcasting in CRNs considering a single-hop scenario, where node *A* is the source node.

- *In traditional single-channel and multi-channel ad hoc networks*, nodes can tune to the same channel for broadcasting. Thus, in Figure 3(a), node *A* only needs one time slot to let all its neighboring nodes receive the broadcast message in an error-free environment.

- *In CRN*, the channel availability is heterogeneous and secondary users are unaware of the available channels of each other. Thus, in Figure 3(b), node *A* may have to use multiple channels for broadcasting and may not be able to finish the broadcast within one time slot.

The exact broadcast delay for all single-hop neighboring nodes to successfully receive the broadcast message in CRNs relies on various factors, e.g., channel availability, number of neighboring nodes and its randomness.

To allow the use of general wireless networks protocols (Figure 3) in CVNs the problems aforementioned must be treated. It is worth to remember that usual broadcasting approaches to CVNs are stateless. In this way, two strategies could be used: (i) a crosslayer design strategy that allows the channel scheduling in broadcasting layer; and (ii) a middleware strategy that considers an additional layer to hide the heterogeneity of channel availability. Song and Xie [49] propose a broadcast protocol, specifically designed to CRNs, that uses similar strategies. The protocol is composed by three components: (i) the construction of the broadcasting sequences; (ii) the distributed broadcast scheduling scheme; and (iii) the broadcast collision avoidance scheme. By using these components they show that the proposed protocol provide successful broadcast ratio while achieving small broadcast delay.

Despite of these issues, research on broadcasting in multi-hop CRNs is still incipient, and consequently it is also in CVNs. There are just few papers addressing the broadcasting issue in CVNs. Two specific protocols are presented in the next subsection the *Coalitional graph games-based algorithm* and *Video streaming over cognitive radio protocol*.

3.1 Coalitional Graph Games-based Algorithm

The *Coalitional Graph Games-Based Algorithm*, proposed by Wang et al. [52], is used to content distribution, i.e., data dissemination, in VANETs scenarios. In this case, a popular file is broadcast by roadside units to vehicles through a particular area. Due to fast speeds and deep fading, some file packets could be lost during the V2I broadcasting stage. Wang et al. propose a P2P approach to allow the vehicles to exchange data and complement the missing packets. The coalitional graph game is used to model the cooperation among vehicles and proposes a coalition formation algorithm to implement the P2P approach.

Authors argue that in VANETs V2V links might be blocked by the factors such as deep channel fading and severe data collisions. By using CRs, it is possible to utilize other available channels with better conditions increasing the aggregate transmission rate and reducing data collisions. In addition, in vehicular environments, particularly on suburban highways, the spectrum is relatively clean, and there are plenty of spectrum holes that can be utilized by CRs. Thus, CRs are beneficial in VANETs.

CRs are utilized for V2V transmissions so that the P2P approach does not require additional bandwidth. The CRs are used to improve the data dissemination when a vehicle is out of range of a roadside unit and the broadcast of the data is already start. Thus, it may consume different unlicensed channels, respecting the primary users priority. Vehicles choose to connect among them in order to enhance its utility. To maximize the utility function, vehicles take into account the channel reliability, it means the effective data rate which was successful received and transmitted. It is defined a probability of successfully transmit a packet assuming a channel with no interference and without both neighbors transmission and primary users activity, simultaneously. In addition, a myopic dynamic algorithm is used, it coordinates the transmission creating a directed graph, and each vehicle has limited information about the network.

3.2 Video Streaming Over Cognitive Radio Protocol

The *VIdeo streaming over COgnitive radio VANETs (ViCoV)*, proposed by Bradai et al. [6], is a protocol for dissemination of safety and entertainment data. It is a video streaming solution that broadcasts safety and entertainment content in both fully and intermittently connected networks under different traffic conditions. The protocol performs in two steps: (i) It selects the best available dedicated or CR channels to disseminate the content; (ii) It chooses a minimum subset of forwarder nodes to reduce the interference achieving a high video quality. CR channels are selected based on their stability over the time, whereas the forwarder nodes are selected based on a new centrality metric inspired from the social network analysis, called dissemination capacity.

Roadside units broadcast the data to vehicles under their ranges. However, the use of roadside units across highway becomes impractical due to high mobility of vehicles, natural interference, and high cost of infrastructure. So, alternatively, ViCoV uses a P2P communication among vehicles with DSRC and CRs. The protocol uses two distinct channels for transition: control channel, to transmit control information; and service channel, to exchange infotainment data. The best available channel is dynamically selected taking into account spectrum utilization, network topology and stability, from DSRC band. However, if these bands have been occupied, the CR approach is used. The channel selection module can operate differently in two scenarios: (i) dense-traffic, where the channel is chosen and a minimum set of neighbors receive the broadcast. This is necessary to avoid broadcast storm; and (ii) sparse traffic, where it is adopted the store-carry forward mechanism.

By using a stable channel, each node selects a set of neighbors to spread its packets. Forwarder nodes are classified by the dissemination capacity metric, based on social networks analysis. Central nodes are chosen due to high dissemination capacity value, it means that packets distributed by these nodes have more chance to reach other nodes with less hops. Due to high traffic and network partitioning a local buffer scheme is used to mitigate the channel overload issue.

4. Neighbors Discovery Algorithms for CVNs

In order to achieve a reliable channel selection in general ad hoc CRNs, ensuring data dissemination accessibility in a non-infrastructured environment, Rehmani [41] discusses about some relevant strategies, where the most important is the neighbor discovery. This approach tries to increase the probability of a message successfully arrives to a receiver node. Other strategies do not considered here, but discussed by Rehmani are: (i) the primary users activity identification; (ii) autonomous decision by CR vehicles; and (iii) the synchronization between nodes sender and receiver.

The neighbor discovery process is a fundamental phase in general CRNs. In this stage, the focus is to determine the neighbors of all nodes in each available channel. This is a non-trivial problem to solve. An effective neighbor discovery algorithm for data dissemination needs to choose the channel which has an appropriated number of neighbors to guarantee the accessibility of the data. This is an interesting challenge because in real environments nodes will be able to operate over multiple channels and each node will probably have different sets of available channels due to distinct spatial location.

Consequently, in CVNs the neighbor discovery is also important. Link disconnections may occur mainly due to high mobility of vehicles on the road. Therefore, identifying a large number of neighbors increases the chance of a packet arrives in its destinations. The most significant feature of CVNs is high mobility, so, CRs strategies are adapted according to fast dynamic changes over time, but it may have an expensive cost.

General strategies to neighbor discovery are based on synchronous or asynchronous operations. A synchronous communication is based on the notion of adjustment among the network elements. Generally, the synchronization process is based on global clock or synchronizer where nodes need to update their local clocks. A specific algorithm to CRN that uses a neighbor discovery strategy is presented by Kondareddy and Agrawal [23], which proposes a channel selection based strategy. Selective broadcasting for multi-hop CRNs is a broadcast algorithm that uses a minimum predefined set of channels, called *Essential Channel Set* (ECS), to transmit information over a limited number of channels. Kondareddy and Agrawal determine the ECS by using two neighbor graphs, one to represent the neighbors and channels where nodes can communicate; and a minimal one that corresponds to its neighbors and the set of channels that nodes can reach all neighbors. The advantage is that it reduces the broadcast delay. The drawback is that it assumes that network topology and channel information of CRs are known. This drawback can hinder the use of this proposal in CVNs.

Additionally, in a synchronous system all nodes begin the neighbor discovery process simultaneously. Due to dynamical geographical location of the nodes, many CRs may have different subset of available channels, to solve this, it is used a deterministic neighbor discovery algorithm for heterogeneous multi-hop CRNs. In this case, besides the neighbor discovery aspect, other important feature is the termination detection. Zeng et al. [53] proposed a neighbor discovery algorithm with termination detection. In their approach, it is guaranteed that all CRs will detect their neighbors and at the end they will terminate the discovery process. The flooding and leader election strategies are used in the neighbor discovery algorithm. The regional leader election process separate node sets in a distributed way considering their distances, so their actions will not interfere in each other. The round robin method is used in neighbor discovery algorithm and executes in the following steps: (i) it divides a round into sub-rounds that correspond to an unique channel, in each sub-round a node compete to be a leader; (ii) it sends its ID to all neighbors, at the end of round robin every node will eventually discover their neighbors; (iii) the node may continue round looking for neighbors which are not identified; and (iv) a termination detection mechanism is used. The termination detection algorithm identifies if there are any node processing the discovery phase. This is performed by flooding an empty message when a node did not finish the last phase. This proposal could be used by CVNs once that the vehicle mobility problem could be treated.

Other examples of synchronous system for neighbor discovery in CRNs are: (i) Jian-zhao et al. [21] proposed a neighbor discovery algorithm where nodes need to be tunned mutually, no previous information about the network is necessary and, due to slotted scheme, all nodes need to start the neighbor discovery process at the same time. It is used a control and data transceiver to improve the performance of the discovery process; and (ii) Mittal et al. [32] present an adaptive algorithm that combine two fast deterministic algorithms for neighborhood discovery. Each algorithm is used for different scenarios. The first one solves the collision-aware problem. It is a fixed slotted algorithm and uses a leader election scheme. The leaders are responsible to start the neighbor discovery process and to inform the nodes about their neighborhood. The second one is used when nodes know the network size. In this case, it uses a gossiping algorithm for a single channel network.

Due to the opportunistic usage of different channels provide by CRs and nodes mobility, the asynchronous neighbor discovery seems to be more adequate to CVNs. An asynchronous communication is done when a node transmits or accesses the channel using different times independently of a global clock. The intrinsic characteristics of asynchronous algorithms increase the complexity of the neighbor discovery process. Specifics asynchronous algorithms to ad hoc CRNs are presented below:

- Bradai et al. [7] propose a distributed channel selection mechanism, called DISCORD, for efficient content dissemination. This mechanism predicts primary user activities by analyzing the channel occupancy and selects the most appropriated channel. It adopts a metric inspired by the *Social Networks Analysis* that calculates a degree for each neighbor to classify the best channels, so, it does not consider the diffusion capacity. In addition, Bradai et al. propose a new metric based on node degree and the shortest path betweenness centrality, called *Dissemination Capacity*. This metric ensures that the 1-hope neighbor has a degree greater than or equal to the local node, so decreasing the chances of a node with a low degree receives a packet.

- Asterjadhi and Zorzi [4] propose the jamming evasive network coding neighbor discovery algorithm, called JENNA. It is concerned not only with neighbor discovery but also with the reliability in the use of channel by CRs. It uses networking code in a single hop network, although each CR node needs to be preconfigured collecting environmental information. This algorithm uses the combination of two approaches, networking code and random channel hopping. Networking code is a technique that creates a linear combination of received and the own packets over the channel, improving the throughput and network security. The problems caused by jamming attacks over RF communications have been gained attention, the random channel hopping scheme is an efficient strategy to solve this issue, since the jammers will not know the hopping sequence due to random behavior. In the neighbor discover phase, CRs are classified into two categories: passive or active mode. In *passive mode* CR hopping randomly collects information about all available channels, e.g., spectrum utilization or QoS, these information will avoid malicious techniques and take information about primary user activity. When the neighbor discovery action is running, the CR assumes the *active mode*. At this time, it schedules control packets and transmits it over free channels. In this phase the number of CRs are estimated along with the identification of presence of reactive jammers.

- Arachchige et al. [3] propose an asynchronous neighbor discovery where a leader election protocol is used in a decentralized network. A CR node is elected a leader based solely on node IDs. This leader becomes responsible to perform the neighbor discovery process and find its neighboring users. The information about neighbors found for each node are also sent. It is unnecessary a global time synchronization, but each node needs to synchronize its own clock ticking when other node sends its clock's value. The algorithm is divided into leader election, neighbor discovery and normal operation phase. In the first stage a CR joins the network, searches for a leader, and, if there is no answer, tries to became one. Neighbor discovery is another important phase, once elected a leader, the node sends a message to CRs allowing other nodes to discover the leader. Ordinary node replies the message and the leader sends an acknowledgment to confirm the inclusion of the node on the network. During the last phase, the leader informs all discovered nodes about their available channel set.

So far, we consider only neighbor discovery strategies (synchronous or asynchronous) in CRNs. Specifically the asynchronous ones could be used in CVNs, since they meet the requirements of VANETs applications. A specific asynchronous algorithm for VANETs, without CRs, that uses a neighbor discovery strategy is presented by Mostajeran et al. [34]. They propose a routing protocol, called I-AODV, that uses neighbor discovery based on flooding strategy. Neighbors of nodes are detected based on periodic HELLO broadcast messages to detect available neighbors at a specific time. Neighbors are recognized and added to a neighbor list after sending an ACK to the sender of the HELLO messages. These HELLO messages are sent only to nodes that are not in the neighbors list. This idea is implemented to decrease possible message overhead in the network. The neighbor node's information is used in flooding route request messages and as a starting point for the data delivery route to destination. This implementation permits the protocol to discover neighbor nodes quickly and uses neighbor node information in the route discovery process. The main issue in considering this approach for CVNs is to allow that, for instance, HELLO messages could be scheduled among available licensed and license-exempt spectrum bands.

Other examples of asynchronous neighbor discovery in VANETs are proposed by Fiore et al. [14] and Kaisser et al. [22]: (i) Fiore et al. propose a lightweight distributed protocol that relies only on information exchange among neighbors, without any need of a priori trustworthy nodes. This protocol is motivated by secure neighbor position discovery problem, an important issue in VANETs; and (ii) Kaisser et al. propose a multi-hop broadcast communication for neighborhood discovery applications which increase driver visibility.

An specific asynchronous algorithm to neighbor discovery in CVNs is SURF [43]. It is a distributed algorithm that can be used by multi-hop CVNs. The algorithm selects a reliable channel on-the-fly, once the channel is chosen, it broadcasts the packets received by neighbors and its own data. Its focus is to disseminate non-urgent messages each hop where the CR tries to find the best channel to delivery its packets to the receiver. There are not routing table nor end-to-end paths, so it is not a routing algorithm. Two features, the largest number of CRs, and most primary user idle times, are used to classify the best channel. The algorithm works in a proactive fashion. Through time observation of the channel primary user unoccupancy, it predicts the idle period of primary user and tries to maximize the chance of receiving a packet by finding the most used channels. All CR nodes work overhearing, even in case of a node does not have any packet to send. In this way CRs act as routers, so, both the sender and the receiver have high chance to be in the same channel, increasing the probability of the data arrives at its destination. Primary user unoccupancy is calculated by using ON/OFF *Markov Renewal Process*, it approximates the spectrum usage through collected historical samples. A busy channel is classified as ON, when there is no primary user activity, and, intuitively, OFF state designates inactivity. Another important feature is a channel learning system once the Markov algorithm does not take into account bad choices. This approach learns with bad channels by saving channel states estimation and observed states, thus, improving future decisions.

High mobility is an important problem in applying CRs, once it traditionally works mostly in stationary nodes. V2V communication is more complex and difficult due to frequent disconnections that occurs when vehicles go out of range. This happens more often when the network density is low. Due to vehicular mobility, the spectrum management could suffer interference as incorrect detection of occupied frequencies happens, mostly caused by node mobility. As we have seen, in VANETs, the basic technique to perform neighbor discovering is to broadcast packets with welcome messages. This idea could be implemented in CVNs with some modifications with a previous scheduling among available licensed and license-exempt spectrum bands.

5. General Data Dissemination Applications in CVNs

There are some research efforts to enable the effective usage of CVNs, but they are still in them preliminary stage. Singh et al. [48] describe some existing projects and testbeds related to CVNs as presented below:

- *Cognitive Radio for Railway Through Dynamic and Opportunistic Spectrum Reuse (CORRIDOR)*[1] is a French research project, finished at October 2014, that targeted opportunistic spectrum access for railways. Communication demands are increasing for modern railways from the railway operations point of view, as well as for providing Internet connectivity to the passengers. However, there is no single universal wireless technology that can answer the needs of heterogeneous services and railway applications. Modern railway applications and passenger's Internet connectivity demand call for more bandwidth and spectrum. Thus, they propose the usage of CRs to support multiple railway applications.

- *The PROTON-PLATA project [17]* is a European project that aims at developing a reconfigurable prototype based on emerging software-defined technologies for telematics applications for V2I and V2V communications. The project proposes a multitechnology cooperative Advanced Driver Assistance System (ADAS) that is based on the integration of Software Defined Radios devices in vehicles.

- *A new cognitive approach to VANETs research* was introduced by Amoroso et al. [2] that enables performing real vehicular experiments, which usually require several vehicles, using only a few vehicles equipped with wireless communication interfaces. This is done by setting up a virtual overlay network consisting of relaying and interfering vehicles on top of a group of only a few vehicles to conduct experiments. The communication packets travel over a random number of hops and experience interference from random number of vehicle transmitters with varying transmission characteristics. A cognitive car testbed [30] is based

---

[1] CORRIDOR—http://corridor.ifsttar.fr. Accessed on April 2015.

on guidelines presented by Amoroso et al. to conduct real experiments with VANETs applications and communication protocols under constraints of vehicular environment and computing resources. The objective of the cognitive car testbed is to explore the possibility of using cognitive network technologies in advanced vehicular testbeds. It focuses on using cognitive radios to study the impact of different frequency bands and different frequency switching delays on the performance of VANETs communication protocols.

These general cases could be used to allow the data dissemination in CVNs. VANETs, equipped with CR or not, are able to collect real-time data on road conditions and make them useful for a wide range of applications [50]. Intelligent transportation system is a widely research field, using cognitive radios approach various applications can be conceived. Some applications, where data dissemination is an important task in VANETs scenarios, are illustrated in the following:

- *Detection of free on-street parking*: In these applications the information about available parking spaces could be disseminated from infrastructure or vehicles to vehicle interested to park. The messages are stored in each intermediate node (infrastructure or vehicles) and forwarded to every encountered node till the destination is reached. Lima et al. [26] perform an analysis of an embedded and ad hoc system to suggest free on-street parking slots by using wireless cameras, without the usage of CRs. The system assumes that the existing cameras deployed in many cities are equipped with an intelligent board capable to communicate and execute the available parking spaces detection system. It is not necessary to install any additional equipment. The parking slots suggestion is performed using image processing techniques. In these case, CR is an efficient solution to take advantage of residual resources of underutilized channels by an opportunistic use of spectrum. Due to the constant use of V2I and V2V communications, this application could be used in a CVN scenario as suggested by PROTON-PLATA project [17], previously discussed. Vehicles can be equipped with Software Defined Radios devices that support different communication scenarios, i.e., V2V and V2I communications, at the same time.

- *Intelligent traffic light*: Smart traffic lights are devices capable of self configuration, able to monitor vehicles flow along the road segments seeking to organize the traffic system and minimize the global wait time of drivers. The traffic information is disseminated to nearest traffic lights in order to improve their inference in the local traffic. Silva et al. [47] address the flow under very low traffic conditions usually found in small cities and during early hours of morning at urban centers. The solution, called LaNPro, avoids situations when the driver has to stop on red light without any vehicle competing for the intersection. LaNPro is a smart traffic light module that schedules vehicles across low traffic intersections. The solution ideally demands wired communication between adjacent traffic lights in order to coordinate the intersections allocation. All vehicles can communicate in an ad hoc fashion to disseminate the traffic lights message. Similar to *Detection of free on-street parking* some CVNs efforts could be used adequately in intelligent traffic light application.

Other VANET data dissemination applications which can execute satisfactory with CRs, i.e., used in CVNs, are listed below:

- *Safety warning systems:* These applications use different alarm message to assist the drivers. A broadcast protocol dynamically chooses a set of nodes to forward the packets in order to avoid collisions and message overhead. The main goal of this protocols is to ensure data accessibility. The alarm messages may be generated, for example, by the activation of airbags or when a crash happens. In both cases, the vehicle starts the broadcast alarm to inform about the danger [5].

- *Vehicle video stream dissemination:* These applications, in general, are based on video broadcast protocols [44]. Several applications and companies use these protocols to transmit safety and infotainment video information. Some examples of video disseminated to the vehicles are: hotel advertising videos, touristic agencies, and highways advertising [6].

- *Vehicle file sharing:* A strategy for peer-to-peer file sharing is proposed by Noshadi et al. [37], and a system for traffic data dissemination was proposed by Nadeem et al. [35] to perform driver assistance and continuous updates about traffic conditions to vehicles.

- *Traffic violation warning:* Kurmar et al. [24] presents a system designed to send warning messages to a driver. It is based on surface and weather condition, as well as speed and distance of vehicle, if it identifies the risk of vehicle pass the red light due to high speed. In addition, services may be implemented as, for instance, *Cooperative Local Applications*. In this service, infotainment applications may be created to obtain local information such as point of interest notifications and local electronic commerce.

- *Medical monitoring:* In the health care field, urgent messages of patient conditions in an ambulance need to be sent on-the-fly to the hospital, so, CRs can be used to establish a communication link over an unused channel and transmit their information over this spectrum. Coordination among traffic lights and ambulances would be interesting in this scenario [38].

6. Final Remarks

Cognitive radio technology is an efficient solution to enhance the spectral utilization, taking advantage of residual resources of underutilized channels by an opportunistic use of spectrum. A cognitive vehicular network allows the cognitive radio to benefit from holes or white spaces presented in the spectrum and admits an opportunistic access that searches for the best wireless communication channel available.

This chapter presented an overview of important technologies used to allow data dissemination in these networks. The topics herein addressed were: (i) *data dissemination and communication*; (ii) *broadcasting*; (iii) *neighbor discovery*; and (iv) *VANETs applications*.

In Data dissemination and communication Section, we discussed different approaches for data dissemination in VANETs: opportunistic data dissemination;

vehicle-assisted data dissemination; and cooperative data dissemination. Additionally, we presented some wireless communication problems in these networks: broadcast storm and security issues.

In Broadcasting Section, we showed the motivation to use broadcasting protocols to disseminate data in these networks. Additionally, we presented two specifics broadcasting protocols: *Coalitional graph games-based algorithm* and *Video streaming over cognitive radio protocol*.

In Neighbors discovery Section, we discussed general strategies to neighbor discovery based on synchronous or asynchronous operations. In the synchronous one we detailed the following algorithms: selective broadcasting; neighbors discovery with lightweight termination detection; fast neighbor discovery; and an adaptive algorithm. In the asynchronous one we discussed the following algorithms: SURF, DISCORD, JENNA, and a Leader Election based.

Finally, in VANETs applications Section, we reported some important application in CVNs context: Detection of free on-street parking, Intelligent traffic light and other general applications. These applications show the importance of data dissemination in CVNs, in special applicability of CRs in real dissemination applications.

References

[1] S. Al-Sultan, Moath M. Al-Doori and Ali H. Al-Bayatti ans Hussien Zedan. A comprehensive survey on vehicular Ad Hoc network. Journal of Network and Computer Applications, 37(2014): 380–392, 2014.

[2] A. Amoroso, G. Marfia, M. Roccetti and G. Pau. Creative testbeds for VANET research: a new methodology. In IEEE Consumer Communications and Networking Conference (CCNC'2012), 2012.

[3] C.J. Liyana Arachchige, S. Venkatesan and N. Mittal. An asynchronous neighbor discovery algorithm for cognitive radio networks. In 3rd IEEE Symposium on New Frontiers in Dynamic Spectrum Access Networks (DySPAN'08), 2008.

[4] A. Asterjadhi and M. Zorzi. JENNA: a jamming evasive network coding neighbor-discovery algorithm for cognitive radio networks. IEEE Wireless Communications, 17(4): 24–32, 2010.

[5] A. Benslimane. Optimized dissemination of alarm messages in vehicular ad-hoc networks (VANET). In High Speed Networks and Multimedia Communications. Lecture Notes in Computer Science, 3079: 655–666. Springer, 2004.

[6] A. Bradai, T. Ahmed and A. Benslimane. ViCoV: Efficient video streaming for cognitive radio VANET. Vehicular Communications, 1(3): 105–122, 2014.

[7] A. Bradai, T. Ahmed and A. Rachedi. Enhancing content dissemination for ad hoc cognitive radio. In International Wireless Communications and Mobile Computing Conference (IWCMC'14), 2014.

[8] M. Chaqfeh, A. Lakas and I. Jawhar. A survey on data dissemination in vehicular ad hoc networks. Vehicular Communications, 1(4): 214–225, 2014.

[9] R. Chen and J.-M. Park. Ensuring trustworthy spectrum sensing in cognitive radio networks. In IEEE Workshop on Networking Technologies for Software Defined Radio (SDR) Networks, 2006.

[10] K. Curran and J. Knox. Disruption Tolerant Networking. Computer and Information Science, 1(1): 69–71, 2008.

[11] S.M.D. Dikaiakos and L.T. Nadeem. VITP: an information transfer protocol for vehicular computing. In ACM International Workshop on Vehicular Ad Hoc Networks (VANET'05), 2005.

[12] B.B. Dubey, N. Chauhan and P. Kumar. A survey on data dissemination techniques used in VANETs. International Journal of Computer Applications, 10(7): 5–10, 2010.

[13] M. Di Felice, R. Doost-Mohammady, K.R. Chowdhury and Luciano Bonon. Smart radios for smart vehicles: cognitive vehicular networks. IEEE Vehicular Technology Magazine, 7(2): 26–33, 2012.

[14] M. Fiore, C. Casetti, C. Chiasserini and P. Papadimitratos. Secure neighbor position discovery in vehicular networks. In IFIP Annual Mediterranean Ad Hoc Networking Workshop (Med-Hoc-Net'11), 2011.

[15] H. Hartenstein and K.P. Laberteaux. A tutorial survey on vehicular ad hoc networks. IEEE Communications Magazine, 46(6): 164–171, 2008.

[16] S. Haykin. Cognitive radio: Brain-empowered wireless communications. IEEE Journal on Selected Areas in Communications, 23(2): 201–220, 2005.

[17] N. Haziza, M. Kassab, R. Knopp, J. Harri, F. Kaltenberger, P. Agostini, M. Berbineau, C. Gransart, J. Besnier, J. Ehrlich and H. Aniss. Multitechnology vehicular cooperative system based on software defined radio (SDR). In International Workshop on Communication Technologies for Vehicles (NET4CARS'13), 2013.

[18] X. He, W. Shi and T. Luo. Survey of cognitive radio VANET. KSII Transactions on Internet and Information Systems, 8(11): 3837–3859, 2014.

[19] Z. Hu, P. Hall and P. Gardner. Reconfigurable dipole-chassis antennas for small terminal MIMO applications. Electronics Letters, 47(17): 953955, 2011.

[20] R. Hussain and M.S. Sharawi. A cognitive radio reconfigurable MIMO and sensing antenna system. IEEE Antennas and Wireless Propagation Letters, 14: 257–260, 2015.

[21] Z. Jian-zhao, Z. Hang-sheng and Y. Fu-qiang. A fast neighbor discovery algorithm for cognitive radio ad hoc networks. In IEEE International Conference on Communication Technology (ICCT'10), 2010.

[22] F. Kaisser, C. Gransart and M. Berbineau. An adaptative broadcast scheme for VANET applications in a high density context. In International Conference on Mobile Ubiquitous Computing, Systems, Services and Technologies (UBICOMM'11), 2011.

[23] Y.R. Kondareddy and P. Agrawal. Selective broadcasting in multi-hop cognitive radio networks. In IEEE Sarnoff Symposium (Sarnoff '08), 2008.

[24] R. Kumar and M. Dave. A review of various VANET data dissemination protocols. International Journal of u- and e-Service, Science and Technology, 5(3): 27–44, 2012.

[25] J.-H. Lim, Z.-J. Jin, C.-W. Song and T.-Y. Yun. Simultaneous frequency and isolation reconfigurable MIMO PIFA using pin diodes. IEEE Transactions on Antennas and Propagation, 60(12): 5939–5946, 2012.

[26] H. David S. Lima, L. Andre L. Aquino, S. Heitor Ramos, Eliana S. Almeida and Joel J.P.C. Rodrigues. OASys: An opportunistic and agile system to detect free on-street parking using intelligent boards embedded in surveillance cameras. Journal of Network and Computer Applications, 46(1): 241–249, 2014.

[27] J. Liu, D. Greene, M. Mosko, J. Reich, Y. Hirokawa, T. Mikami and T. Takebayashi. Using utility and micro utility for information dissemination in vehicle ad hoc networks. In IEEE Intelligent Vehicles Symposium (IV'08), 2008.

[28] D.F. Macedo, S. de Oliveira, F.A. Teixeira, A.L.L. Aquino and R.R. Oliveira. (CIA)2-ITS: Interconnecting mobile and ubiquitous devices for intelligent transportation systems. In IEEE International Conference on Pervasive Computing and Communication (PER-COM'02), 2012.

[29] G. Maia, A. Boukerche, A.L.L. Aquino, A.C. Viana and A.A.F. Loureiro. A data dissemination protocol for urban vehicular ad hoc networks with extreme traffic conditions. In IEEE International Conference on Communications (ICC'13), 2013.

[30] G. Marfia, M. Roccetti, A. Amoroso, M. Gerla, G. Pau and J.-H. Lim. Cognitive cars: constructing a cognitive playground for VANET research testbeds. In International Conference on Cognitive Radio and Advanced Spectrum Management (COGART'11), 2011.

[31] C.N. Mathur and K.P. Subbalakshmi. Digital signatures for centralized DSA networks. In IEEE Consumer Communications and Networking Conference (CCNC'07), 2007.

[32] N. Mittal, S. Krishnamurthy, R. Chandrasekaran, S. Venkatesan and Y. Zeng. On neighbor discovery in cognitive radio networks. Journal of Parallel and Distributed Computing, 69(7): 623–637, 2009.

[33] A. Mor. Study of different type of data dissemination strategy in VANET. International Journal of Engineering Science and Innovative Technology, 1(2): 6–8, 2012.

[34] E. Mostajeran, R. Md Noor and H. Keshavarz. A novel improved neighbor discovery method for an Intelligent-AODV in Mobile Ad hoc Networks. In International Conference of Information and Communication Technology (ICOICT'13), 2013.

[35] T. Nadeem, P. Shankar and L. Iftode. A comparative study of data dissemination models for VANETs. In International Conference on Mobile and Ubiquitous Systems: Networking & Services (MOBIQUITOUS'06), 2006.

[36] S.-Y. Ni, Y.-C. Tseng, Y.-S. Chen and J.-P. Sheu. The broadcast storm problem in a mobile ad hoc network. Wireless Networks, 8(2-3): 153–167, 2002.

[37] H. Noshadi, E. Giordano, H. Hagopian, G. Pau, M. Gerla and M. Sarrafzadeh. Remote medical monitoring through vehicular ad hoc network. In IEEE Vehicular Technology Conference VTC'08, 2008.

[38] P. Phunchongharn, E. Hossain, D. Niyato and S. Camorlinga. A cognitive radio system for e-health applications in a hospital environment. IEEE Wireless Communications, 17(1): 20–28, 2010.

[39] Md. J. Piran, Y. Cho, J. Yun, A. Ali and D.Y. Suh. Cognitive radio-based vehicular ad hoc and sensor networks. International Journal of Distributed Sensor Networks, 2014(154193): 11, 2014.

[40] S. K.M. Rao and A.V. Reddy. Data dissemination in mobile computing environment. BVICAM's International Journal of Information Technology, 1(1): 57–60, 2009.

[41] M.H. Rehmani. Opportunistic data dissemination in ad-hoc cognitive radio networks. Ph.D. thesis, Université Pierre et Marie Curie—Paris VI, 2011.

[42] M.H. Rehmani and Y. Faheem. Evolution of cognitive networks and self-adaptive communication systems, chapter Data Dissemination and Channel Selection in Cognitive Radio Networks, pp. 30–48. IGI Global, 2004.

[43] M.H. Rehmani, A.C. Viana, H. Khalife and S. Fdida. SURF: A distributed channel selection strategy for data dissemination in multi-hop cognitive radio networks. Computer Communications, 36(10-11): 1172–1185, 2013.

[44] C. Rezende, A. Boukerche, H.S. Ramos and A.A.F. Loureiro. A reactive and scalable unicast solution for video streaming over vanets. IEEE Transactions on Computers, 64(3): 614–626, 2015.

[45] C. Schroth, R. Lasowski and M. Strassberger. Vehicular Networks, chapter Data Dissemination in Vehicular Networks, pp. 181–221. Auerbach, 2009.

[46] R. Shivanagu and C. Deepti. A security mechanism against reactive jammer attack in wireless sensor networks using trigger identification service. International Journal of Security, Privacy and Trust Management, 2(2): 43–54, 2013.

[47] C.M. Silva, L. Andre, L. Aquino and W. Meira-Jr. Smart traffic light for low traffic conditions: A solution for improving the drivers safety. Mobile Networks and Applications, 2015(1): 11, 2015.

[48] K.D. Singh, P. Rawat and J.-M. Bonnin. Cognitive radio for vehicular ad hoc networks (CR-VANETs): approaches and challenges. EURASIP Journal on Wireless Communications and Networking, 2014(49), 2014.

[49] Y. Song and J. Xie. A distributed broadcast protocol in multi-hop cognitive radio ad hoc networks without a common control channel. In IEEE INFOCOM (INFOCOM'12), 2012.

[50] Y. Toor, P. Muhlethaler, A. Laouit and A. de la Fortelle. Vehicle ad hoc networks: Applications and related technical issues. IEEE Communications Surveys & Tutorials, 10(3): 74–88, 2008.

[51] A.M. Vegni, M. Biagi and R. Cusani. Vehicular Technologies—Deployment and Applications, chapter Smart Vehicles, Technologies and Main Applications in Vehicular Ad hoc Networks, pp. 3–20. InTech, 2013.

[52] T.Wang., L. Song and Z. Han. Coalitional graph games for popular content distribution in cognitive radio VANETs. IEEE Transactions on Vehicular Technology, 62(8): 4010–4019, 2013.

[53] Y. Zeng, N. Mittal, S. Venkatesan and R. Chandrasekaran. Fast neighbor discovery with lightweight termination detection in heterogeneous cognitive radio networks. In International Symposium on Parallel and Distributed Computing (ISPDC'10), 2010.

[54] X. Zhang, W. Cheng and H. Zhang. Heterogeneous statistical QoS provisioning over 5G mobile wireless networks. IEEE Network, 28(6): 46–53, 2014.

[55] J. Zhao and G. Cao. VADD: Vehicle-assisted data delivery in vehicular ad hoc networks. IEEE Transaction Vehicular Technology, 57(3): 1910–1922, 2008.

[56] J. Zhao, Y. Zhang and G. Cao. Data pouring and buffering on the road: A new data dissemination paradigm for vehicular ad hoc networks. IEEE Transaction on Vehicular Technology, 56(6): 3266–3276, 2007.

5

Routing Protocols for Cognitive Vehicular Ad Hoc Networks

Wooseong Kim[1],* and Mario Gerla[2],*

ABSTRACT

The need for V2V communications in future vehicles will escalate because of driver demands for content, especially as vehicles are becoming more autonomous, thus allowing drivers to download files and streams from the Internet. Since LTE and DSRC spectrum is scarce, vehicles will need to tap the unlicensed ISM spectrum to establish V2V channels in order to communicate and download. However, the WiFi spectrum is overcrowded with residential users requiring channel-agile, cognitive radios to mitigate interference. In this chapter, we introduce the concept of CoVanet, a Cognitive Vanet that features cognitive radios specifically designed for the urban scenario. The CoVanet features are applied to CoRoute, an anypath routing protocol that trades off robustness (to primary interference) with path length and exploits multiple path diversity. Cognitive Multicast, or CoCast, is then introduced for popular applications downloading. CoCast offers innovative features such as parallel data block transmissions in OFDM subchannels and network coding across parallel channels. CoRoute and CoCast QualNet simulation results indicate that they outperform existing wireless routing schemes in interference prone urban environments.

1. Introduction

This chapter explains multi-hop routing protocols, unicast and multicast protocols, for cognitive vehicular ad hoc networks (VANET). Dedicated Short Range Communications architecture (DSRC) for the vehicular communications has been standardized and

---
[1] Computer Engineering, Gachon University.
[2] Computer Science, University of California, Los Angeles, CA, USA.
* Corresponding authors: wooseong.kim@gmail.com; gerla@cs.ucla.edu

deployed at 5.9 GHz designated spectrum, but it has narrow bandwidth and limited number of channels that are mostly reserved to safety applications. VANET applications other than safety, especially high data rate media transmissions, are supposed to use other radio resources such as 3G, LTE and, of course, Wi-Fi. Wi-Fi is attractive because it is unlicensed and has many channels across a wide spectrum. However, deploying the VANETs using the Wi-Fi in dense urban areas is challenging due to interference and coexistence with residential equipment that also operates in the unlicensed bands.

In this chapter, we present a new architecture, **Cognitive Multi-hop Ad Hoc Network** in the unlicensed bands for the urban VANETs. This architecture borrows concepts from flexible and cognitive radio platforms to develop high capacity multiple channel and multiple radio ad hoc networks. In this network, two routing protocols for unicast and multicast are introduced, respectively. The new unicast routing protocol named *Cognitive Ad hoc Vehicular Routing Protocol (CoRoute)*, exploits channel state (as proactive routing schemes such as link state do) and geo-location information (as stateless GeoRouting protocols do), simultaneously. This is a new vehicular unicast routing protocol that accounts for external interference from road side residential Wi-Fi access points. A *Cognitive on-demand Multicast Routing Protocol (CoCast)* establishes multicast trees on multiple channels jointly with a novel channel allocation algorithm based on the spectrum sensing. This is the first multicast routing protocol adopting the cognitive radio technology.

2. Cognitive Vehicular Ad Hoc Networks in Unlicensed Bands

In the new cognitive VANET architecture operated on unlicensed bands, vehicles exploit multiple channels to increase throughput by avoiding interference from residential 802.11a/b/g access points along the road. We redefine the terms used in conventional cognitive radio to clarify difference between conventional cognitive radio and our approach. "Primary Nodes" (denoted as PNs) are external nodes that are not cognitive and do not belong to the ad hoc network—the secondary network (i.e., the VANET). For example, residential Wi-Fi access points along the road in Figure 1 are the "Primary Nodes" and are external to the VANET. "Cognitive Nodes" (denoted by CNs) are the mobile ad hoc nodes, i.e., the secondary nodes that have the ability to scavenge dynamically the channels least occupied by the primary users, PNs. Note that this architecture is different from conventional cognitive radio systems in frequency, network topology and primary user protection. For instance, vehicles as secondary nodes do not need to vacate frequency bands immediately to defer to primary traffic because they are operating in unlicensed band instead of, say, licensed TV band. Our model reduces the impact caused by external interference to PNs in order to achieve robustness and high performance exploiting spectrum sensing and following Dynamic Spectrum Access (DSA) principles.

This section overviews CoVanet, a novel Vanet multi-radio multi-channel cross-layer architecture based on cognitive radio principles.

2.1 CoVanet Network Architecture

Figure 1 illustrates a simple CoVanet architecture. Residential 802.11 APs are located along the road (typically inside building) and use various IEEE 802.11 channels as shown in Figure 1. Due to varying transmission power and shadowing, radio coverage and interference vary from an AP to an AP so that each road segment has different interfering channels and interference signal power levels. For example, in Figure 1, channel 1 and 6 suffer interference from APs in the road segment S1 while only channel 1 suffers from interference in road segment S4. Furthermore, in some portion of S1 there is interference on channel 1, but other portions of S1 have interference on channels 1 and 6. Therefore, vehicles experience constantly changing channel interference while moving on the road.

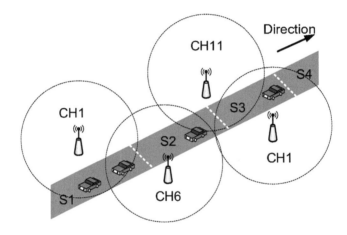

Figure 1. Model of cognitive vehicular ad hoc networks.

Vehicles in CoVanet are equipped with multiple radios. In this study, we assume each vehicle has two radios (i.e., R1 and R2), which allow vehicles to transmit and receive packets simultaneously via R1 and R2 respectively. We assume that R1 is tuned to the node's receive channel while R2 is transmitting. Typically, the channel with lowest interference is selected for receive interface, R1. Therefore, the R1 receive channel changes as the vehicle travels through the urban grid. R2 on the other hand is dynamically switched to frequency hopped Common Control Channel (CCH) and, if there is a packet to transmit, it is tuned to a receive channel of the next hop to which the packet must be transmitted. In the current hardware platforms, switching delay is assumed to be 100 us while packet transmission time is 5–6 ms in 2-Mpbs data rate radio. The two radio approach achieves adaptive channel diversity while maintaining stable connectivity. With a single radio, the vehicles would require a separate rendezvous mechanism to tune their transmitter and receiver to the same channel before exchanging a packet.

2.2 Common Control Channel Hopping

CoVanet requires a robust CCH for exchanging control messages. If CCH was based on a preselected ISM-band channel, it would become a bottleneck due to congestion or interference from PNs or from other CNs. To overcome this problem, nodes in CoVanet use a frequency hopped CCH that pseudo-randomly spans all the channels. Common channel time synchronization for the entire VANET is obtained from the GPS signal with timing error less than 1 *usec*. Frequency Hopping period and sequence are predefined and known to all vehicles.

Nodes broadcast Hello messages to neighbors periodically, say every few seconds, to exchange information about network topology, current receive channel for R1, channel utilization, position, etc. The transmit radio R2 interface jumps to CCH periodically to transmit Hello messages to neighbor nodes after each data transmission. Thus, the R2 is always tuned to the CCH unless occupied with a transmission. Typically, a node selects its R1 receive channel before sending a Hello message on the CCH to help maintain local connectivity. To avoid collisions, random jitter is inserted before transmitting a Hello control message on the CCH channel.

2.3 Distributed Channel Assignment Algorithm

Vehicles as CNs in CoVanet monitor a randomly picked subset of the receive channels and share the monitored information with neighbor nodes in a collaborative sensing process. This guarantees that all channels are monitored even if each CN is tuned to only a few channels, at the same time limiting sensing overhead and spreading it uniformly over all CN's. In order to monitor PN traffic load in isolation without the CN load contribution, CoVanet employs a quiet period synchronized across the VANET, during which all CN nodes stop data transmission and listen to PN's traffic on the randomly picked channels. The quiet period is set to 20 ms each second. The CN node senses PN radio states: busy or idle.

PN channel workload (ω_i) is estimated for each channel from quiet period monitoring and is propagated to all neighbors. From the measured PN workload ω_i and from the physical data rate W_0 (e.g., 11 Mbps), the expected residual capacity of channel i, W_i (i.e., the capacity available to secondary users) is calculated as:

$$W_i = W_0 (1 - \omega_i)$$

Each CN radio also monitors its own received traffic W' in a proper history window (say a few seconds) and shares its own window averaged traffic (i.e., of received packets x average packet size) with neighbors through CCH. Each CN then approximates the residual channel capacity available on channel i as:

$$W''_i = W_i - W'_i$$

where W'_i is a sum of averaged traffic rate of all neighbors that monitored channel i. Each node CN selects as its new receive channel, j, for the next time period the channel with highest residual capacity (and thus least congestion) as following:

$$j = arg \max_i W''_i$$

Then, it informs the neighbors of its new channel frequency in the next Hello message.

3. CoRoute: A New Cognitive Anypath Vehicular Routing Protocol

In this chapter, we introduce the Cognitive Ad hoc Vehicular Routing Protocol (CoRoute), namely, is a hybrid vehicular routing protocol that exploits channel state and geo-location information. Existing geographic based routing protocols, such as GPSR, GPCR and GPSRJ+, are suitable to support robust connectivity with relatively low overhead even in high vehicle mobility. They, however, do not account for spectrum limitations and for interference and conflicts. This is the first attempt to VANET routing with cognitive radios that achieves both scalability and link quality (i.e., delay) stability, adjusting to high mobility as well as spectrum scarcity.

3.1 CoRoute Design

Existing VANET routing protocols aim at robust connections with low connectivity maintenance in presence of high vehicle mobility. In dense urban areas, heavy channel interference by residential users is a critical factor that must be considered in selecting the route.

This CoRoute selects low interference channels and explores low interference paths. The following principles have guided CoRoute design:

- *Multiple connections between vehicles:* In MANETs network connectivity improves with node density. In urban environments, however, density alone is not sufficient. Connectivity can change rapidly and unpredictably due to obstacles, traffic signals, etc. Connections are intermittently lost, for example, when neighbors are blocked by other vehicles or buildings. To maintain connectivity in a VANET, regardless of density, it is not sufficient to exchange Hello messages to detect changes in a specific topology. One must keep track of multiple possible alternative topologies.

- *Next hop robustness:* Link quality in CoVanet is determined by many factors. The urban environment because of buildings, structures, and hills introduces multipath fading with diffraction and scattering. Mobility and speed cause Doppler spread. Finally, PN interference from residential APs impacts channel quality. Thus, each node must dynamically select the next hop that exhibits the most robust channel. Instead of maintaining a single best next link (and changing it when network conditions change), we use the concept of best forwarding set of links/neighbors and opportunistically let the most favorable next hop prevail.

- *Connectivity vs. Congestion:* High vehicle density and robust connectivity do not guarantee better throughput because of potential data traffic congestion. Each vehicle must consider data network congestion generated by PNs and CNs.

Figure 2 illustrates the outcome of the CoRoute protocol. A source node S searches for the shortest path to destination D. There are two candidate routes, the

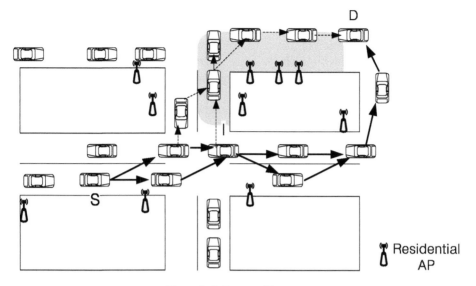

Figure 2. CoRoute architecture.

upper route (dashed line) and the lower route (solid line). Both have same length (i.e., same number of hops). CoRoute selects the *lower route* because of *lower interference*.

Due to varying channel environment and topology, CoRoute does not implement a proactive algorithm with global link state information as in link-state routing (e.g., OSPF, OLSR, etc.) or with constant background exchange of route state information like Distance Vector. Rather, it implements a hybrid protocol called "ETX/ETT assisted GeoRouting".

We borrow from published literature the Expected Transmission Time (ETT) [1] and Expected Transmission count (ETX) [2] methods, a typical choice for delay metric in wireless ad hoc networks. More precisely, we use ETT as it better reflects packet loss and channel bandwidth. The cumulated ETT along the path is used to account for path quality (i.e., shortest path). Also we combine this approach with anypath forwarding. The anypath paradigm forwards packets to multiple forwarding candidates in parallel. This characteristic makes anypath very robust against interference caused by PNs.

For the GeoRouting, each node should know its own position, the position of its neighbors and the position of the destination node. How positions are obtained or shared is outside the scope of this paper. We assume that each node is able to obtain its own position using GPS devices, exchange it with neighboring nodes by control messages via CCH, and obtain the position of the destination node by a separate location service (see [3] for a summary of location services). Initially, the next hop to destination is selected mainly using GeoRouting since there is no available end to end ETT value. However, the criterion for route selection among the set of candidate "greedy" GeoRouting neighbors is influenced by the ETT metric. As the packet progresses to its destination, path state is created at the intermediate nodes. Namely, the delay estimate ETT is propagated from the current node along the path back to the source. When the packet reaches the destination, the source has the estimated delay to destination along the georouted (i.e., minimum geographic distance) path as well as on alternate paths. Basically, as packets are routed from source to destination, the routing scheme evolves from GeoRouting to Minimum Delay routing based on the cumulated ETT metric. The ETT metric is updated as conditions change and the path is modified accordingly. The ETT shortest path method also helps to overcome a critical problem in GeoRouting, namely dead-ends. If in the initial phase GeoRouting leads to a trap node, an infinite cumulative delay value, e.g., cumulated ETT = 10000 seconds, is propagated back to the source, forcing new packets to follow another path, until a breakthrough, if any, is found. The above hybrid routing strategy combines the scalability of GeoRouting and the end to end delay optimality and deadend prevention of Distance Vector routing.

To illustrate this scheme, consider Figure 2. A packet in node 'I' must be routed to 'D'. The neighbor in the upper road is closer to the destination. However, the lower road is selected because the delay is lower. In the next section, we discuss in more detail the computation of ETX and ETT estimates and their use as our routing metrics.

3.2 ETX Estimation

ETX is the average number of transmissions required to deliver a packet across a link. It can be measured by broadcasting probing packets at a very slow rate and averaging the received number of broadcast probes during a predefined window. Let p_f and p_r be the packet loss probabilities in forward and reverse directions, respectively. Then, under the assumption of unlimited retransmissions, the probability of transmission failure and ETX are respectively,

$$p = 1 - (1 - p_f)(1 - p_r)$$

$$ETX = \sum_{k=1}^{\infty} kp^{k-1}(1 - p) = \frac{1}{1 - p}$$

Using probe packets to measure ETX introduces considerable overhead in a multi-channel network like CoVanets since each node must probe multiple channels. Yet, the ETX estimate is essential to guide the GeoRouting algorithm to find good paths, as outlined in the previous section. To avoid this overhead, in CoRoute, each node estimates ETX using a probabilistic channel model based on measured spectrum sensing information, on estimated PN workload and on measured distance between a sender and a receiver. ETX is then used for ETT computation and link delay prediction, as shown in the above equation. It is assumed that all vehicles CoVanet know their location via GPS. Loss probability is computed under fixed data rate between a pair of vehicles with uniform transmission power, and for given propagation characteristics and modulation assumptions.

3.3 Primary Interference Packet Loss

We use here a simplified model based on the rectilinear layout in Figure 3. A collision occurs when the PN transmitter is "hidden" from the CN transmitter as shown in

Figure 3. Thus, whenever the PN and CN signals coexist, the CN packet is lost by the receiver. Two cases are considered. The optimistic model, called capture model, assumes that vehicle V2 receives correctly from V1 as long as the distance between V2 and V1 is smaller than the distance between V2 and T1, where T1 is the PN node position (in Figure 3, an Access Point). The pessimistic model assumes that vehicle V2 fails to receive any time when the distance between V2 to T1 is <= d, where d is the transmission range (for simplicity we assume that the transmit power is uniform over all PNs and CNs).

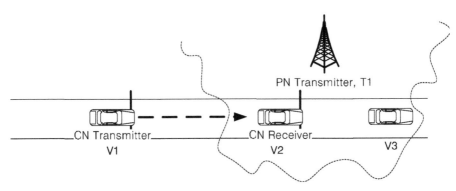

Figure 3. PN hidden terminal from CN transmitter on the road.

Referring to Figure 4, in the pessimistic model any partial overlap of the CN and PN signals at the receiver will cause failure. In the optimistic model, failure occurs only if the PN signal is stronger than the CN signal. We make here the "conservative" assumption that the AP does not implement the IEEE 802.11 RTS/CTS option. In fact, even if the CN and PN transmitters are hidden from each other, RTS/CTS option in the PN would eliminate the collision. We also make the assumption that the PN and CN packets are of the same size. This assumption is not critical for this study, but it simplifies the model. Consider Figure 3, and assume that the distance between V1 and V2 (=d). In both conservative and optimistic model, V2 suffers a loss when the distance between V2 and T1 is <= d. Recalling that the PN load in the hearing range (−d, d) is ω, say, node V2 suffers a loss with probability $P_c = \omega/2$ if PN is in the interval (0, d). However, recall that the collision occurs either if a CN transmits first, and a PN starts its transmission later, or; the PN comes on first and the CN starts to send later. This is the well known "unslotted Aloha" model that leads to double the collision loss probability with respect to the slotted Aloha model (we assume that arrivals are Poisson and packet sizes are identical and exponentially distributed). Thus $P_c = \omega$. Now assume that the distance (V1, V2) decreases. With the optimistic, capture model, we note that the loss probability becomes 0 when (V1, V2) = d/2. In fact, because of CN capture, no PN can interfere with S2 unless it also came within range of V1 (which would violate the hidden terminal conditions). The fraction of PNs that interfere with V2 decreases linearly as the distance (V1, V2) decreases. Thus, averaging over all (V1, V2) relative positions we obtain the average $P_c = \omega/2$. Considering now the worst case (no capture), we note that the PN will disruptively interfere with V2 until (V1, V2)

= 0 and in this case only the PN at the distance d will interfere, with zero probability as the size of the PN interfering set is null. Thus, the collision loss probability in this case decreases linearly from [V2 at the distance d from V1] to [V2 superimposed with V]. The collision loss probability averaged over the entire range of possible (V1, V2) configurations in case of no capture (worst case) is thus $P_{c,wrs} = \omega$. It is worth noting that upon successful reception, the receiver responds immediately with an ACK. The transmitter is awaiting for this ACK. The ACK reception is protected from PN interference and collision because the potential colliders have been deferred by CN's transmission. Moreover, the ACK packet is much shorter than the data packet, thus loss due to noise is negligible. Based on these considerations, we assume that the P_r, the loss probability in the reverse direction is 0.

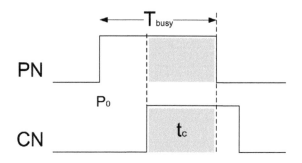

Figure 4. Collision between a PN and a CN.

3.4 Model Validation

In Figure 3 we did not consider the possibility of an intersection between V2 and T1. We note that the intersection marginally reduces the fraction of PN interferers along the rectilinear direction (less building space for residential users) until node V2 is itself in the intersection, at which point the interference significantly increases (while the CN is in the intersection) because V1 is exposed to the PN users in the cross street. The probability of a receiver to be in the intersection is higher in the grid topology than in the linear topology, since in the grid topology the nodes in the intersection are frequent forwarders because they can lead to alternate paths, as confirmed by the experiments in the evaluation section. In summary, the actual P_c value will range between $\omega/2$ and ω. There could be arguments about using the conservative as well as the optimistic model. At this time we will use the conservative, worst case model, leaving the trade off between the two models to future work.

Figure 5 depicts a simulation result from Qualnet about CN's packet loss rate (P_c) due to collision with PN traffic. The simulator calculates the packet error using collision duration and interference strength based on a SINR/BER table that is created by empirically measured data. A CN transmitter sends packets with varying packet sizes from 400 to 1400 bytes to a CN receiver that has a hidden PN transmitter as shown in Figure 3. At the same time, the hidden PN generates channel workload from 0.2 to 0.9 by shaping traffic patterns with varying T_{idle} from 1 to 20 ms and fixed

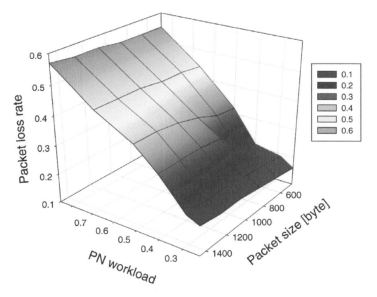

Figure 5. Packet error rate due to PN collisions under varying PN traffic and CN packet size.

T_{busy} = 6 ms that is the approximated time to transmit a 1500 byte packet using 2-Mbps data rate. From the simulation result, we can verify that packet error rate to calculate the ETX ranges between the aforementioned boundaries, $\omega/2$ and ω. From the $P_{d(d, k)}$ and P_c the loss probability of forward transmission can be calculated as $p_f = 1-(P_{d(d, k)} (1-P_c))$ in a receiver node for MAC DATA. Here recall that p_r is negligible. While the ETX estimate is not very accurate (because of the various approximations introduced in the model), it is still valuable for the purpose of guiding the GeoRoute search onto low delay paths. It should be noted that when the path is established and traffic flows on it, the nodes along the path actually measure ETX from packet transmissions. Thus, the delay metrics for paths in use are the measured metrics, and the minimum delay arguments are valid for these paths.

3.5 ETT Estimation

Expected transmission time (ETT) is the transmission time to deliver a packet with size S and data rate W over a link. Namely:

$$ETT = ETX \frac{S}{W}$$

where S is packet size, ETX is the ETX value and W is the effective data rate on the link in question. Available data rate from PN workload (ω) is $W = W_0(1 - \omega)$ when both PNs and CNs can share the channel bandwidth by avoiding mutual collision using 802.11 MAC protocol collision avoidance. Namely, the effective rate W reflects the waiting time of the packet in the buffer due to the preemption by other stations during count down. Loss and retransmission that are caused by random errors or by collisions (e.g., hidden terminal situations) have been already factored in ETX. As a result, a route with heavy PN traffic will result to have higher ETT and will be avoided by CoRoute.

3.6 Anypath Routing

In CoRoute, vehicles select a route "on demand" among multiple candidates based on estimated channel conditions. CoRoute uses a hybrid scheme that was earlier defined as ETX/ETT assisted, Cognitive GeoRouting. The scheme is inspired by GeoRouting [4] and by "ETX AnyPath Routing" [5]. Like in [4] CoRoute greedily looks for the two-hop neighbor that makes the most progress to the destination. The two hop neighbor information is contained in the Hello messages. It was shown in [4] that GeoRouting via two-hop neighbors greatly improves performance at intersections. CoRoute follows the opportunistic anypath routing based on the forwarding set technique first introduced in [5]. The key idea is that instead of unicasting the packet to the next hop, the packet is broadcast to a forwarding set in order to exploit spatial redundancy. This technique is especially useful in a scenario where there is external interference caused by PNs as the probability that one node in the forwarding set should receive the packet correctly is higher than sending just to a single next hop. In [5] vehicles in the forwarding set are priority ranked. Lower priority nodes transmit a packet after time out, if they do not hear a transmission from higher priority nodes. Else, they discard the packet to suppress duplicated forwarding. Node priority is determined by the end to end delay, Expected Anypath Transmission Time (EATT).

In [5], EATT is the delay to the final destination and it is computed using a proactive, global link state algorithm. In CoVanet, EATT initially is the delay to the selected two-hop neighbor that offers the best progress to the destination. In fact, the selection of the best 2-hop node is based both on geographic progress and expected EATT. This combined metric is important as it bans 2-hop neighbors with infinite EATT, since they lead to a trap. The process is shown in Figure 6. The EATT choice is motivated by the fact that anypath routing guarantees the minimum delay route from a source to a destination [5].

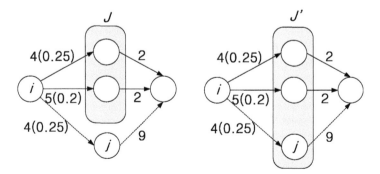

Figure 6. The forwarding set of any path routing. The link labels represent EATT (ETX). The forwarding set at the left-hand side figure excludes node *j*; the one at the right-hand side figure includes node *j*. The addition of node *j* leads to a HIGHER cumulative EATT, thus node *j* is NOT included in the forwarding set.

Initially our algorithm minimizes EATT to the 2-hop target. However, as packets reach the destination, the EATT values are propagated from the destination back to the source through periodic Hello messages. After that happens, a generic intermediate node j will, like in [5], send packets on the path that minimizes EATT estimate to the destination.

Figure 6 illustrates how ETT values (d_j) and success transmission probability ($p_j = 1/ETX$) denoted in parenthesis for each link are used to calculate EATT, i.e., D_i based on the Bellman equation (i.e., $D_i = d_{i,J} + D_J$) across forwarding set J. Since the d_j is proportional to the number of transmissions, d_j is determined by $1/p_j$. More precisely:

$$d_{i,J} = \frac{1}{1 - \prod_{j \in J} 1 - p_j} \frac{S}{R_j}$$

where R_j is defined in W''_i, D_J is calculated as weighted sum of each contribution across the forwarding set

$$D_J = \sum_{j=J} \omega_j D_j, \ where \ \sum_{j \in J} \omega_j = 1$$

where ω_j is the probability that forwarder j is successfully used, and is given by:

$$\omega_j = \frac{p_j \prod_{k=1}^{j-1} 1 - p_k}{1 - \prod_{j \in J} 1 - p_j}$$

As an illustration, in Figure 6 using the above formula we find that the forwarding set J has average 4.5 EATT(= 1/(1–(3/4)(4/5)) + (2/4+(3/4)(1/5)*2)/(1–(3/4)(4/5))) when we assume that the average bit duration S/R is 1. As per definition in [5] the forwarding set consists of the minimum number of forwarders that creates the lowest EATT based on above equations. The forward set J' in the right figure has EATT = 5.7, which implies that the additional forwarder j increases EATT and thus should not be included in the set based on the forwarding set definition in [5]. The D_i is updated as Hello messages convey fresh information from the destination. Thus, the forwarding set adapts dynamically to the topology and PN traffic. As mentioned earlier, the EATT feature assists CoRoute to recover from dead ends, a common problem in GeoRouting. When a local maximum (in terms of progress to the destination) is reached, CoRoute drops the packet and propagates an infinite EATT information towards the source. This assures that future packets in the session are routed via a different neighbor that does NOT lead to the same trap. Packet drop makes most sense in real time streaming scenarios as we consider in this case. Other recovery schemes could also be used if reliable packet delivery is required. For instance, the packet could be routed back to the source, for retransmission. Or, the packet could be carried by the vehicle towards its destination in a Delay Tolerant fashion, as described in [4]. If the application is time sensitive, and vehicles have a 3G radio (as is usually the case), the packet can be transmitted to the destination via 3G. This is an expensive (as 3G is not free) but necessary solution in emergencies. It should be pointed out that the original forwarding model in [5] was based on a single channel configuration. In our case the model was modified to support multiple channels. Thus, the cumulative EATT along the path corresponds to link EATT values computed for different channels. The forwarding also differs from the simple broadcast used in [5]. Here multiple broadcasts on different frequencies are required for different subsets of forwarding nodes. Due to multiple broadcasts to forwarding nodes on different channels, and due to the fact that additional delays are introduced by the use of the CCH channel (to prioritize the forwarders), the claim of optimality (with respect to average end to end delay) reported in [5] must be carefully revisited in this scenario to identify the restrictions under which optimality still applies.

3.7 Forward Set Implementation

In CoRoute, each vehicle uses Hello messages on the CCH in order to propagate to all neighbors its local perceived status including sensed channel information, two-hop geographical location, etc. This information is used by each node to select the least congested receive channel on which to listen to. It is also used to select the two-hop target and the associated forwarding set. Namely, based on the neighbor table, the vehicle with a packet to send selects the 2-hop neighbor geographically closest to the destination and forms the forwarding set J.

A node transmits the packet to the forwarding set J using multiple broadcasts on multiple channels. To avoid unnecessary duplicate forwarding, each node in J has a priority that is based on its EATT to the two-hop target. Namely, the time out is proportional to EATT. To suppress superfluous multiple transmissions, the top priority node upon successful forwarding broadcasts a "notification" message on the CCH. The CCH notification is required (in addition to ACK and echo ACK) because some of the forwarders may be on a different channel than the sender. If a forwarder on a different channel misses the CCH notification, it may transmit a duplicate packet, which will be filtered out by the target node.

3.8 CoRoute Evaluation

We implemented CoRoute and compared it with other routing schemes using Qualnet 3.9.5. The simulation platform runs IEEE 802.11b with 2-Mbps data rate and 11 orthogonal channels with Rayleigh fading. The mobility traces for the simulations were generated using VanetMobiSim. Both macro and micro mobility are accounted for in the vehicular environment so as to reproduce a realistic urban motion pattern. A 1500 m x 1500 m Manhattan grid with 300 m road segments was used. All roads have a speed limit of 15 m/s (54 km/h). The micro-mobility is controlled by the Intelligent Driver Model (IDM-IM). The number of primary interfering PNs (i.e., residential Access Points) ranges from 50 to 300. These nodes are placed randomly in the urban grid. All simulations were run with a fixed, single source-destination session with a 64 Kbps UDP data traffic transfer. This single vehicular session assumption implies that there is no contention among secondary nodes (i.e., vehicles) unless they are in the same session.

To assess the benefits of cognitive radios, we compare first CoRoute with the GeoRouting scheme Route, which is identical to CoRoute except for the use of only one radio tuned to a single channel. Next, we compare CoRoute to an AODV based variant. The reason for choosing AODV is that many of the previous cognitive radio routing protocols were based on a cognitive AODV approach. For realism we have developed a modification of CoRoute called CoAODV. This allows us to compare CoRoute with a representative example of the class of prior AODV based cognitive routing systems. CoAODV is identical to AODV, except it runs on a cognitive radio platform and exploits multiple channels. CoAODV achieves path optimization using cumulative EATT like CoRoute. However, since the channel environment and the network topology change dynamically, the path must be refreshed accordingly. We note that CoAODV uses only one path, thus it cannot rely on the redundancy of the forwarding set as CoRoute does. Instead, it must reflood the route request (RREQ) message periodically to refresh the route and reselect the channels. Because of this refresh, CoAODV does not require Hello messages like CoRoute. In our experiments, the refresh period is set to 5 seconds. As expected, the flood overhead introduces significant additional overhead and negatively affects CoAODV performance in very congested (i.e., heavy PN load) conditions.

3.8.1 Convoy(Chain) Topology

We start with a set of experiments where the cars move in a convoy, as a motorcade snaking through urban streets. There is high vehicle density in the convoy, thus the "convoy topology" guarantees end to end connectivity and allows us to ignore disconnection problems. For the convoy topology, we evaluated the performance of CoRoute for different numbers of interfering primary stations PNs; and for different PN utilization factors.

Figure 7 shows CoRoute average packet delivery ratio (PDR) for variable number of PNs and for three levels of PNs loads (i.e., 20, 40 and 70%). The PNs are added incrementally reflecting a monotonic increase from 50 to 300 PNs. A single source-destination pair moving along the same path is randomly chosen, and it remains the same for the entire set of experiments. As expected, PDR decreases monotonically with the number of PNs as interference increases and available bandwidth becomes scarce. The PDR also decreases with increasing primary occupancy. Figure 8 shows

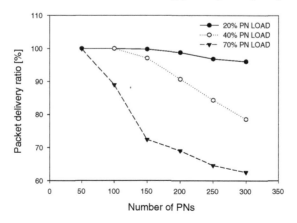

Figure 7. Packet delivery ratio with varying PN workloads.

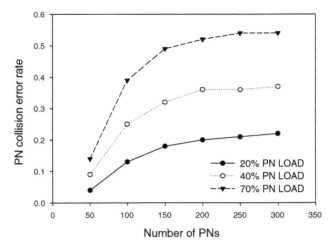

Figure 8. Loss rate due to PN collision under varying PN workloads.

the packet loss rate due to PN collisions. PN collision rate in CoRoute increases with the number of PNs and with PN load, as expected.

In order to investigate effect from the any path routing, we compare normal single path routing (SPR) and any path routing (APR) of CoRoute. Both schemes use cognitive radio approach with multiple channels. Figure 9 shows average packet delivery ratio (PDR) of them with variable number of flows in the same convoy topology where 50 PNs are deployed with 70%. In the figure, CoRoute outperforms the single path routing by 20% in case of 6 flows due to redundant packets caused by the any-path forwarding procedure.

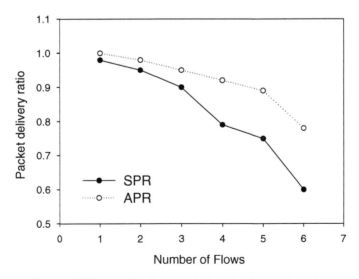

Figure 9. PDR comparison between single path and any path routing.

Figure 10 shows overhead with ratio of the number of source CBR packets against all transmitted packets from all nodes. The overhead increases exponentially according to number of flows, and overhead in APR of CoRoute is higher than SPR due to redundant packets from imperfect ACK procedure in CCH. In the SPR case, the number of source packets of 6 flows is 5% to all packets in the networks. But the APR of CoRoute is 2.5% which is one half of the SPR; it is almost double to the SPR. From this result,we conclude that the overhead of the CoRoute can limit network throughput especially in high node density and large number of flows.

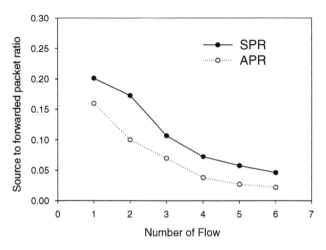

Figure 10. Overhead comparison between single path and any path routing.

The second set of experiments compares CoRoute with Route. PDR are shown in Figure 11. As in the previous experiment, there is a single source-destination pair and it is the same in all the runs. PNs are 70% loaded and are added incrementally. The

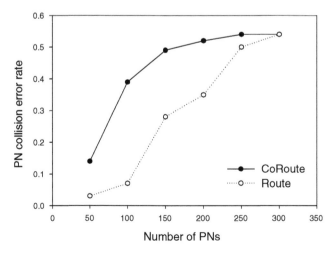

Figure 11. Error ratio due to PN collisions in CoRoute and Route.

PDR decreases monotonically as the number of PNs increases. CoRoute out performs Route by a margin of 20 to 70% over the entire range due to multi-channel diversity.

Figure 12 shows the loss rate resulting from PN collisions. Both schemes suffer of monotonically increasing PN collision losses as the number of PNs increases. Interestingly, PN collision rate in CoRoute represents a larger fraction than Route loss rate. This is because Route suffers more losses from other factors like intra-flow interference while CoRoute avoids it by using different channels hop by hop.

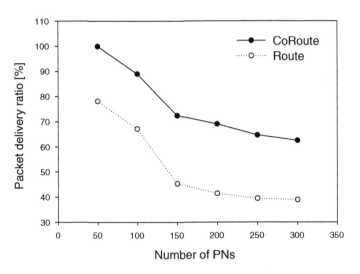

Figure 12. PDR comparison between CoRoute and Route.

3.8.2 Grid Topology

The next series of experiments features CoRoute in the grid (i.e., Manhattan) topology. The grid topology is clearly more realistic than the convoy(chain) topology. Since vehicles now move on multiple paths, the network is exposed to more frequent disconnections.

This effect is visible in Figure 13, which shows that the packet delivery ratio(PDR) in the grid where 100 vehicles are deployed, for the same PN density, is about 2 to 3 times lower than for the convoy.

The impact of the vehicle topology configuration on connectivity is also clearly visible in Figure 14 where the fraction of packets lost due to lack of connectivity is shown to be twice as high for the grid as for the convoy for the same PN density. This result also indicates the PN interference is more considerable in grid topology rather than convoy since the proportion of packet loss increases drastically in grid topology.

We evaluate more about the sensitivity of PDR to vehicular density and its overhead with a single flow in CoRoute. In this experiment, we select a random source and destination pair in contrast to previous simulation that uses a fixed pair for all runs. Only 50 PNs are deployed to reduce impact from PN interference. Figure 15 shows that average PDR with 95% confidence interval increases monotonically according to number of vehicles.

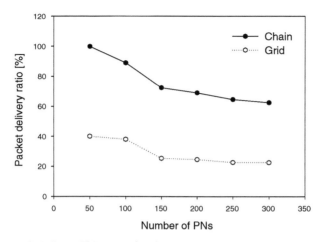

Figure 13. CoRoute PDR comparison between convoy(chain) and grid topology.

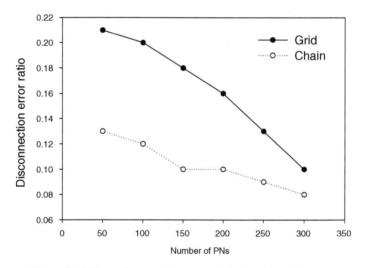

Figure 14. CoRoute disconnection error ratio in the grid topology.

Figure 16 shows overhead from variable vehicular density in grid topology. The overhead that is denoted by packet ratio between number of source packets and all transmitted packets for the flow in an entire network increases exponentially according to number of vehicles, since larger number of vehicles produce more duplicated packets although it helps increasing PDR.

In Figure 17 we show a comparison between CoRoute and Route delivery ratios for 60 and 100 vehicles respectively in grid topology. Here, a fixed single source-destination pair is used for all the runs. The horizontal axis shows the number of PNs, as usual. CoRoute delivery performance improves by about 25% as the number of vehicles is increased from 60 to 100. This is due to the topology connectivity increase. In contrast, Route performance decreases by about 5 to 10% as the number of vehicles

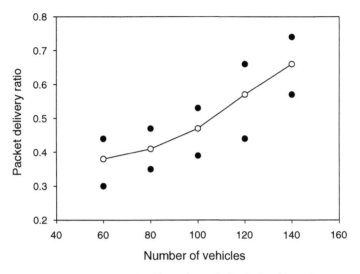

Figure 15. CoRoute PDR with varying node density in grid topology.

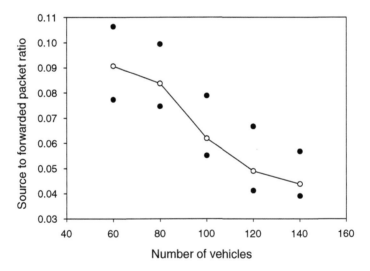

Figure 16. CoRoute overhead in grid topology; ratio of number of source packets to all transmitted packets in networks.

increases. Apparently the high vehicular density causes more collisions in the Route scheme, which operates on a single channel. This offsets the gains brought about by increased connectivity. PDR of CoRoute in Figure decreases drastically rather than one in the 100 vehicles since the pair of a source and destination experiences more interference in 60 vehicle case. From this, the PDR in 60 vehicles looks converging into one of Route while the PDR in 100 vehicles is almost in parallel.

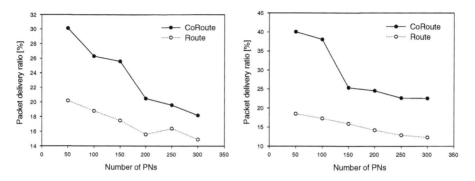

Figure 17. PDR comparison between CoRoute and Route in 60 and 100 vehicles.

Figure 18 shows aggregate delivery ratio of CoRoute and Route protocols averaged over several experiments with PN number varying from 10 to 200; 40% workloads; different, randomly picked communication pairs, and; varying number of vehicles ranging from 60 to 120. Here again CoRoute outperforms Route exhibiting a PDR of 60% as compared to 35% for Route. This firmly establishes the superiority of cognitive radios in congested spectrum environments. The figure also shows the performance of CoAODV and AODV, which will be discussed in the next section.

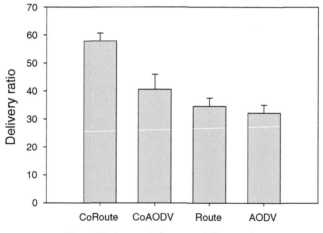

Figure 18. Aggregated average delivery ratio.

As mentioned earlier, most of the previous cognitive radio routing protocols were based on AODV. For fairness, we compare here CoRoute to CoAODV. The first experiment is based on the convoy topology. In Figure 19 we report PDRs for variable PN density and high PN utilization (at 70%). At light PN density, i.e., with PN number from 50 to 100, CoAODV does much worse than CoRoute. This is because CoAODV must refresh the path via flooding to overcome the frequent changes. CoRoute is not

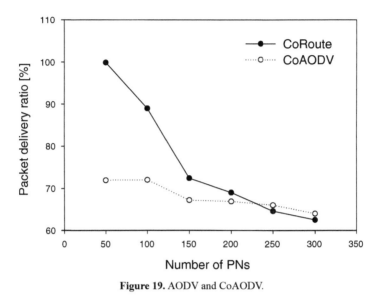

Figure 19. AODV and CoAODV.

so sensitive to changes because of the forwarding set redundancy. At moderate PN load, say above 200 PNs, CoRoute starts having problems locating a feasible path, while CoAODV at no extra cost finds the feasible path if there is one. Consequently, at heavy PN load, CoAODV does slightly better than CoRoute.

In this chapter, we proposed a novel routing scheme, CoRoute, for cognitive VANETs. CoRoute achieves two goals. It makes best use of the available Wi-Fi bandwidth and it causes minimum disruption on residential users. CoRoute is inspired to state of the art VANET routing protocols. It is based on GeoRouting, the most popular scheme in VANETs. Moreover, it exploits the ETX delay metric and the forwarding set concept proved optimal in Mesh networks. CoRoute consistently outperforms conventional GeoRouting schemes (by almost 100%) as well as AODV based cognitive routing schemes such as CoAODV, yielding in this case up to 25% throughput improvement.

4. CoCast—Multicast Mobile Ad Hoc Network Using Cognitive Radio

CoCast consists of vehicles and Wi-Fi access points as primary nodes (PNs) that potentially interfere with each other in an urban environment as shown in Figure 20. CoCast attempts to provide adequate throughput performance to vehicles by selecting and using least congested channels. Like before, each vehicle is equipped with two radio interfaces (i.e., R1 and R2) that are tuned to different channels to exploit spectral diversity and used to receive and transmit packets simultaneously. Here channels of each interface R1 and R2 are assigned by a hybrid channel assignment algorithm [6] in which a channel for the receiving interface R1 does not change for a long time (e.g., every 10 minutes) but a channel for the transmitting interface R2 is able to be changed dynamically among the channels used by neighboring vehicles for their

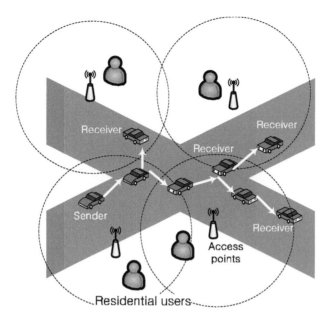

Figure 20. CoCast architecture.

receiving. So, there is no extra overhead required for channel rendezvous mechanism between intended sender and receiver since a receiving interface is semi permanently tuned to its preferred idle channel.

4.1 Multicast Route Establishment and Channel Selection

In order to achieve an optimal channel assignment for the receiving radio interface R1, vehicles sense spectrum periodically on both interfaces and estimate PN traffic workload. Following radio resource measurements based on channel occupancy and/ or interference [8], vehicles then estimate PN traffic workload, ω, on each channel. Synchronization for sensing the PN traffics among the vehicles could be realized by using GPS with timing error less than 1 μs [9]. The longer the sensing window, the better the workload estimation, but the more overhead the sensing will take. Collaborative sensing that exchanges measured channel information among the vehicles also helps reducing the overhead.

After measuring the residual channel capacity from PNs, nodes collaborate to form a multicast tree using an ODMRP [14] type technique and jointly select their own channels. When a source has packets to send, it starts broadcasting a Join Query (JQ) packet via a predefined common channel. Upon receiving the non-duplicated JQ packet, a node records the upstream node address in the routing table to learn the reverse path and rebroadcasts it. When the JQ reaches a multicast member node, the node creates a Join Reply (JR) packet and sends it toward the source. The JR packet is relayed through the learned reverse path and the nodes on the reverse path are organized as a "forwarding group". After exchanging the JQ/JR, the multicast route is created and data is delivered through this multicast route. During the route establishment period, CoCast executes channel allocation. JQ/JR contain channel information (i.e., a selected channel and its PN workload) as shown in Figure 21. More precisely, the JQ contains the candidate channel list (L_c) based on spectrum sensing (e.g., 1, 2 and 3 on a JQ message in the figure). Upon receiving the candidate channel list from an upstream node, the node selects a listening channel from the received channel list. To select the listening channel, the node compares the received channel list with its own available channels. And the node rebroadcasts the JQ after removing the upstream node's candidate channels and writing its own candidate channel list. Upon receiving the JQ, a receiver node writes the selected listening channel on the JR and sends it toward the source.

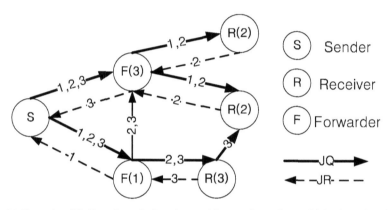

Figure 21. Example of CoCast channel allocation; multicast nodes assign multiple channels using Join Query (JQ) and Joint Reply (JR). Number in each node indicates a selected channel.

Sensed channel information is updated periodically using JQ-JR messages; the multicast tree is refreshed periodically (every few seconds) to adjust to network topology and channel status. All control messages are broadcasted through the CCH. CCH can be established by GPS. For instance, all vehicles with a same channel hopping sequence meet on a certain channel using the R2 interface every CoCast refresh interval (e.g., 4 seconds) and beaconing interval (e.g., 1 second).

4.2 Enhanced CoCast: Combat ACI with OFDM Subchannelization and Network Coding

CoCast using the partially or fully overlapped channels can experience throughput reduction due to interference. We mitigate such ACI effect using the mechanisms detailed in the remainder of this section.

Packet Transmission in OFDM Subchannels

Our protocol groups multiple subcarriers into a logical unit called a subchannel. An example is shown in Figure 22, where 8 subcarriers are grouped into a subchannel, resulting in a total of 6 subchannels (i.e., 48 subcarriers) that form an individual OFDM channel.

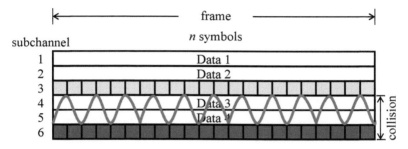

Figure 22. Four data blocks embedded in an IEEE 802.11a/g OFDM frame. Data 3 and 4 are lost due to ACI.

A packet from the upper layer arriving at the MAC layer is further split into data blocks, or simply blocks and each block is assigned to a different subchannel, after proper physical-layer processing that includes channel coding, interleaving, and scrambling. This way, multiple data blocks from the original packet are transmitted in parallel, over multiple subchannels. Compared to the scheme where the entire data frame is spread over all m subcarriers, this scheme is more robust against narrowband interference and ACI. Namely, the blocks on some subchannel will still survive even when the blocks in other subchannels are completely obliterated by excessive interference. Furthermore, if the transmitter can predict the amount of interference that is expected in a specific subchannel, it can dynamically select the subchannels with no change required to the physical layer structure.

In the example in Figure 22, 4 data blocks are sent in subchannels 1, 2, 4, and 5 in parallel. Suppose data in subchannels 4 and 5 experience interference from the adjacent channel. If ACI causes bit errors that cannot be corrected by forward error correction (FEC), the entire frame will be dropped. Recall that in the standard 802.11a/g, the frame spans the whole channel (that is, all the 48 subcarriers). In contrast, in our scheme, the blocks transmitted in subchannels 1 and 2 survive under the interference.

4.2.1 Mitigating the Effect of ACI

Given that parallel transmission of data blocks over multiple subchannels is enabled, one can think of several techniques to combat ACI. The conventional approach is to sense the spectrum beforehand, and use only the subchannels that are unlikely to be interfered. Since such prediction can never be perfect, a recovery scheme is often devised to ensure reliable delivery of data. "Remap" [10] was proposed to permute the data block-to-subchannel mapping in subsequent packets, so that different data blocks are affected in consecutive packet retransmissions (see Figure 23). This way, the receiving node can extract and combine from the repeated transmissions the blocks that survived the interference thus drastically reducing the number of retransmissions.

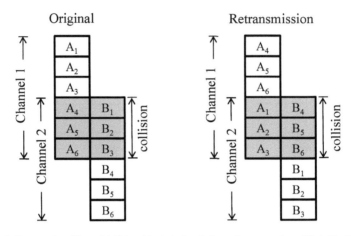

Figure 23. In Remap, the effect of ACI is mitigated via subchannel permutation of data blocks (A1 to A6) in retransmissions. Here, the simultaneous transmissions of frames A and B experience ACI, yet can be recovered successfully after two transmissions.

Unfortunately, such a retransmission scheme is not applicable in CoCast—**no retransmission** is allowed in multicast in order to avoid packet implosion. Also, we cannot expect that a receiver can receive enough randomly permutated data blocks to recover from multiple forwarders of a mesh structure to acquire same gain because the multicast topology in CoCast is more tree-like than mesh-like due to multi-channel usage.

In addition, figuring out the optimal way to permute data blocks at each transmission is not a trivial problem. In the example of Figure 23, a receiver needs four transmissions to recover the packet in the worst case. The number of transmissions is not even predictable if interference in the subchannels rapidly fluctuates in time. Instead of Remap, for CoCast we propose to use network coding to mitigate the effect of ACI. More precisely, the data blocks in a frame are network coded using random linear coding (RLC) and the coded blocks are transmitted over the parallel subchannels. First, the added advantage of performing parallel transmissions in CoCast is the exploitation of the spectral diversity of subchannel interference where narrow band fading and ACI are selectively present. Secondly, with network coding, permutations of data blocks are not necessary. The receiver can decode successfully when it obtains k correct, linearly independent coded blocks where k is the number of blocks in the frame. For example, suppose each frame is divided into $k = 6$ blocks and these blocks are encoded as shown in Figure 24.

Assuming that half of the subchannels are affected by ACI, two transmissions are sufficient for the receiver to collect enough blocks (i.e., 6 coded blocks) to decode the entire frame. In this particular example, a node receives the same frame from

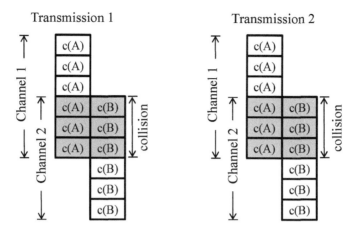

Figure 24. In our protocol, blocks are network coded. Let c(m) = c1m1 + c2m2 + ~...~ + ckmk, where the coefficients c1, ~...~, ck are randomly selected for each block and k is a size of generation vector for network coding. Here the generation is the aggregate of the blocks sent to all subchannels. The receiver needs to collect k blocks.

two different forwarding nodes in the mesh network; the two forwarders tune to a same receive channel. Thus, all nodes can receive packets through the transmitting interface to increase receiving opportunity while it is not used for transmitting. If this redundancy is not sufficient to recover an original packet (i.e., no mesh redundancy because there is a single forwarding node), this node will dynamically adjust the coding rate to increase redundancy to a proper level, for example, a factor of 2 based on overheard receiver activities [11]. In this example, for each frame that is divided into k blocks, the single node generates a total of $2k$ coded blocks. The higher the coding rate and the redundancy, the higher the reliability. However, this comes at the cost of a possible throughput reduction.

4.3 CoCast Evaluation

We have implemented CoCast with the multi-radio and the OFDM subchannel model in QualNet simulator [15]. Each node is equipped with two radio interfaces; one is for receiving packets and the other is for transmitting packets [12]. The receiver scans the channels to measure primary nodes (PNs) workload every second during a synchronized quiet period. When it detects that the workload of PN interferers in its current receive channel exceeds ones of other channels by a predefined threshold, it switches to another receive channel. After then, it notifies the neighbors of the channel change via beacon or through the next JQ-JR message exchange in the multicast tree refresh procedure. Recall, the default beaconing and refresh interval are 1 and 4 seconds, respectively.

   Residential traffic is modeled by several interference generators representing individual PNs with a two-state (busy and idle) continuous Markov model having randomly generated workload (i.e., busy probability). The workload is referred to as interference rate [12].

4.3.1 Freeway Topology

We demonstrate feasibility of CoCast in a freeway, 50 m × 1500 m, as shown in Figure 25 where topology does not change frequently since the grouped vehicles move in same direction and speed, thus disconnection does not happen if vehicle density is adequate. Along the route of freeway, 10 ~ 200 PNs are deployed, evenly distributed over the 11 channels. The interference rate is set to 60% for every interference generator (unless stated otherwise). We assign each multicast source a data flow rate = 400 Kbps, with 6 Mbps channel rate.

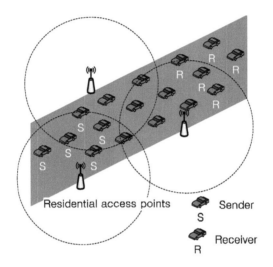

Figure 25. Freeway simulation model. Five multicast source nodes, S, share the same set of members as receivers denoted by R.

4.3.2 CoCast versus ODMRP

First, we compare original ODMRP and CoCast in terms of packet delivery ratio (PDR) under varying number of multicast flows and PNs in the freeway. Note that the original ODMRP was implemented using a single channel and is shown as case 1-channel in Figure 26 in which the radio is simply tuned to the same channel in all nodes. In contrast, the multiple channel options that enables the CoCast's capability of scavenging idle channels are cases 3 and 11-channel respectively.

We simulate multiple flow scenarios without any PN interference to test ODMRP scalability. As can be seen in Figure 26(a), packet delivery ratio (PDR) decreases drastically especially in the single channel case (i.e., ODMRP) as the number of multicast flows from different sources increases. Multiple channels help reduce collisions from the multiple streams, but PDR is different depending on number of available channels; 11 orthogonal channels gain 30% performance over 3 channels. Next, Figure 26(b) depicts a single flow PDR with varying number of PNs near the multicast group. The single channel suffers serious packet loss due to PN interference while the 11 channels sustain PDR. Here we can see that the cognitive ability of CoCast helps maintaining adequate performance in the face of PN interference.

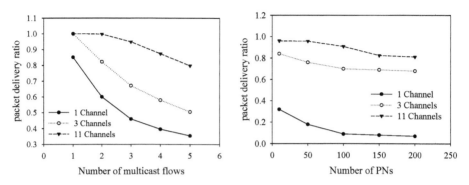

Figure 26. Throughput comparison with multiple flows and PN interference in a freeway simulation model.

4.3.3 CoCast with Network Coding

In this section, we investigate the network coding effect on multicast in the VANET as a solution to increase PDR. Although CodeCast [13] already showed improved PDR of multicast using random linear network coding in a MANET, in this simulation we capture the network coding effect in multiple channels (i.e., CoCast) as opposed to single channel (i.e., CodeCast). The redundancy factor for network coding (i.e., coding rate) is fixed at 3; a source node generates 3 times as coded packets as original packets for a multicast stream at the rate of 400 Kbps.

Figure 27 plots multicast performance with varying network coding rate in the single channel. Network coding rate helps increase PDR by an eight-fold wrt no coding (to PDR = 65%) for 200 PNs. However, higher redundancy than coding rate = 4 leads to throughput reduction due to congestion. This network coding gain also costs end-to-end delay as shown in Figure 27. Note that the delay with any coding rate is much higher than no coding (i.e., coding rate = 0). However, delays with coding rate = 3, 4

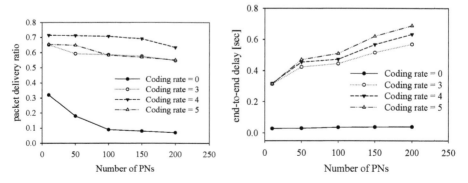

Figure 27. Network coding effect on single channel.

and 5 are comparable, implying that the delay from network coding accounts for most of the delay, while collisions from redundant packets play a secondary role. In this simulation, we configure generation size of network coding as 8 where nodes should wait 8 packets to generate 8 coded packets. The single channel with network coding does not combat the PN interference completely compared to multiple channel; low coding rate is not enough to achieve 100% PDR and high coding rate causes congestion.

Figure 28 shows CoCast throughput when applying various network coding rates. Unfortunately, CoCast does not acquire high benefit from network coding due to diversified receiving channels of vehicles which lessens receiving opportunity from neighbor nodes. Instead, the transmitting interface can catch coded packets on another channel but they are not as many as the single channel case. Low coding rate can drop PDR lower than no coding because a single coded packet loss can lead to the loss of 8 consecutive packets. Thus coding rate = 1 in Figure 28 plummets drastically at 100 PNs while coding rate = 3 shows almost similar throughput with no coding case. The network coding for CoCast also comes at the cost of a high increase in the end-to-end delay, as shown in Figure 28. Compared to the single channel network coding, delay is consistent regardless of number of PNs due to cognitive radio capability. However, delay at small number of PNs is little bit longer than the single channel due to the channel switching and scanning overhead required for the cognitive operation in the CoCast.

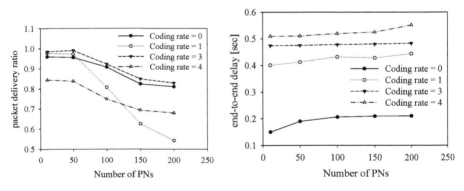

Figure 28. Network coding effect on CoCast with multiple channels.

4.3.4 CoCast with OFDM Subchannels

In previous subsections, the multiple channels used for CoCast are assumed orthogonal. However, there are only 3 true orthogonal channels in the 2.4-GHz ISM band. Simulation results show that 3 channels are not enough to avoid intra/inter flow interference and external PN interference. Thus we use partially overlapped 11 channels. In order to investigate the overlapped channel effect, in presence of Network Coding, we implemented the partially overlapped channel model described in Qualnet; the ACI effect is given according to the spectral mask defined in the 802.11 standards.

Figure 29 shows the effect of ACI and transmission type (serial or parallel) in terms of packet delivery ratio (PDR) and end-to-end delay of each case. Figure (a) plots the PDR of CoCast for ACI OFF and then ON with varying redundancy rate. In a real multichannel environment where ACI is present, PDR suffers severe packet loss especially with high redundancy = 3 since increasing redundancy leads to additional packet loss on other overlapped channels. The result shows that 11 channels with ACI is almost comparable with orthogonal 3 channels in terms of PDR as shown in the Figure (b).

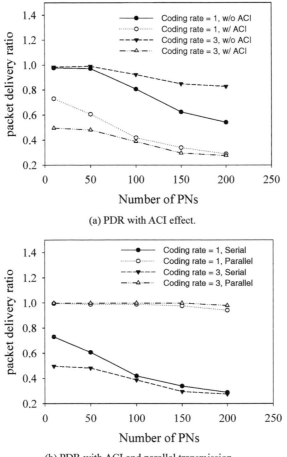

(a) PDR with ACI effect.

(b) PDR with ACI and parallel transmission.

Figure 29. Enhanced CoCast with OFDM subchannels to combat ACI effect.

Therefore we try to develop the enhanced CoCast with parallel transmission using OFDM subchannels to combat such ACI. In the channel environment, we implemented 6 OFDM subchannels by dividing a single 802.11 channel to achieve our proposed idea, ACI reduction, in which a packet is also divided into 6 data blocks and they are encoded by random linear coding. Each coded block is delivered through each subchannel.

We recall that in the serial transmission mode, each frame occupies the entire OFDM channel in the conventional way, while in the parallel mode, the frame is subdivided into blocks which are sent over different subchannels in parallel. As can be seen in Figure 29 PDR with coding rate = 3 reaches almost 100% in the parallel modes. With ACI ON, all cases in the parallel mode outperform the serial mode, which confirms that they are well protected from ACI.

Figure 30 depicts enhanced CoCast performance with varying redundancy in partially overlapped 11 channels. Network coding rate = 3 shows almost 100% PDR in Figure 30. Surprisingly, the case of coding rate = 1 also manages to achieve comparable PDR even with no redundancy, thanks to parallel transmissions and to the implicit redundancy of network coding. In case of no coding, however, the advantage of parallel transmission is not as significant as in coding cases. Note that delay of no coding is comparable to the coding cases since parallel transmission causes the same delay to gather all data blocks as network coding in Figure 30. From this we note that the delay of network coding cases is almost the same as in serial transmission CoCast. Moreover delay with coding rate = 3 is lower than for rate = 1 in Figure 30 since waiting time for all coded blocks (i.e., 8 blocks) in case of coding rate = 3 is much lower than 1.

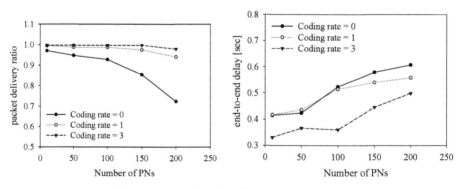

Figure 30. Network coding effect on enhanced CoCast.

4.3.5 Manhattan Grid Topology

We evaluate the enhanced CoCast in a 1500 m x 1500 m Manhattan grid topology with 300 m road segments. The mobility traces for the simulations were generated using VanetMobiSim. Both macro and micro mobility are accounted for in the vehicular environment so as to reproduce a realistic urban motion pattern. All roads have a speed limit of 15 m/s (54 km/h). The micro-mobility is controlled by the Intelligent Driver Model (IDM-IM). 60 vehicles are deployed randomly within the simulation space. One sender and five receivers are selected as multicast members and are randomly distributed in the grid. Multicast flow rate is 200 Kbps from the sender to the receivers; 11 channels with realistic 802.11 channel model are used for the vehicles and PNs. PNs are given 60% workload and are also uniformly distributed in the space and channels.

Figure 31 shows the enhanced CoCast throughput in the Manhattan grid where vehicular mobility is diversified and multicast members are scattered over the entire space. As shown in the Figure, network disconnection from the mobility and multi-hops reduces PDR heavily by 40% compared to the freeway scenario. However, high redundancy with coding rate = 3 is consistent in showing that network coding is effective for high mobility networks. Delay of all cases also increases compared to the freeway where vehicles are located in line of sight and covered by small number of hops.

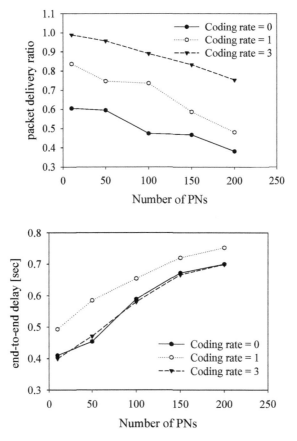

Figure 31. Enhanced CoCast performance in Manhattan grid.

5. Summary

The need for V2V communications in future vehicles will escalate because of driver demands for content, especially as vehicles are becoming more autonomous, thus allowing drivers to download files and streams from the Internet. However, direct content downloading from the Internet from Access Points and from LTE towers is becoming difficult because of the competition of residential users (for the former) and the spectrum limitation (for the latter). Moreover, the DSRC spectrum will be by and large reserved for safety applications. Thus, vehicles will need to tap the unlicensed ISM spectrum to establish V2V channels so that they can extend the LTE and the (rare) WiFi downloading to the crowds. Since the WiFi spectrum is overcrowded and the residential users have first rights to it, the use of channels agile, cognitive radios is required. In this chapter, we introduce the concept of CoVanet, a Cognitive Vanet that features cognitive radios on board and pursues two goals: efficient ISM spectrum scavenging and; deferral to Residential users (viewed here as Primary users). The key features of the CoVanet radios are first described, including: the Control Channel implementation using Frequency Hopping, to reduce spectrum O/H and simplify rendez vous, and; the channel assignment algorithm that gives priority to Residential users.

Next, the CoVanet features are applied to the first protocol, CoRoute, an anypath routing protocol that trades off robustness (to primary interference) with path length. CoRoute innovative feature is to exploit the diversity offered by multiple paths (as reflected in the qualification of "anypath routing". CoRoute is shown to outperform all current Cognitive Vanet implementations.

While multi-hop file downloading and uploading will be an important requirement for non popular files, the downloading of popular content and the dissemination of emergency information to vehicles will use multicast routing. To this end, we have extended the cognitive radio concept to multicast. Cognitive Multicast, in brief, CoCast, is cognitive multicast protocol especially designed for VANETs. Beyond the services of conventional multicast such as ODMRP, CoCast offers innovative features that make it robust to ACI and frequency selective fading, namely parallel data block transmissions in OFDM subchannels and network coding across parallel channels. The CoCast protocol has been implemented and evaluated in the QualNet simulator. Simulation results indicate that CoCast outperforms existing wireless multicast schemes in an interference prone environment. More specifically, parallel block transmissions and network coding help avoid external interference effectively, especially in a channel environment where channels overlap in frequency.

References

[1] R. Draves, J. Padhye and B. Zill. Comparison of routing metrics for static multihop wireless network. Proc. of ACM SIGCOMM, 2004.

[2] D.S.J. Couto, D. Aguayo, J. Bicket and R. Morris. A high-throughput path metric for multi-hop wireless routing. Proc. of ACM MobiCom, 2003.

[3] J. Wilcox, T. Camp and J. Boleng. Location information services in mobile ad hoc networks. In IEEE International Conference on Communications, 2002.

[4] K. Lee, J. Haerri, U. Lee and M. Gerla. Enhanced perimeter routing for geographic forwarding protocols in urban vehicular scenarios. Proc. of IEEE Globecom Workshops, 2007.

[5] Rafael Laufer, Henri Dubois-Ferriere and Leonard Kleinrock. Multirate anypath routing in wireless mesh networks. In Proc. of IEEE Infocom, 2009.

[6] P. Kyasanur and N. Vaidya. Routing and link-layer protocols for multi-channel multi-interface ad hoc wireless networks. ACM SIGMOBILE MC2R, 10(1): 31–43, 2006.

[8] Stefan Mangold and Lars Berlemann. IEEE 802.11k: Improving confidence in radio resource measurements. Proc. of IEEE PIMRC, 2005.

[9] Peter H. Dana. Global positioning system (GPS) time dissemination for real-time applications. Real-Time Systems, 12: 9–40, 1997.

[10] Li Erran Li, Kun Tan, Ying Xu, Harish Visvanathan and Y. Richard Yang. Remap decoding: Simple retransmission permutation can resolve overlapping channel collisions. Proc. of ACM MobiCom, 2010.

[11] Soon Y. Oh, Eun-Kyu Lee and Mario Gerla. Adaptive forwarding rate control for network coding in tactical manets. Proc. of IEEE MILCOM, 2010.
[qualnet] Qualnet 4.5. http://www.scalable-networks.com/products/qualnet/.

[12] W. Kim, A. Kassler, M. Di Felice and M. Gerla. Urban-x: Towards distributed channel assignment in cognitive multi-radio mesh networks. Proc. of IFIP Wireless days, 2010.

[13] Joon-Sang Park, M. Gerla, Desmond Lun, Yunjung Yi and M. Medard. Codecast: a network-coding-based ad hoc multicast protocol. IEEE Wireless Communications, 13(5): 76–81, 2006.

[14] Yi, Yunjung, et al. On-demand multicast routing protocol (ODMRP) for ad hoc networks. draft-yi-manet-odmrp-00. txt (2003).

[15] http://web.scalable-networks.com/content/qualnet

6

Routing in Cognitive Vehicular Networks

Ozgur Ergul[1,*] and *Ozgur B. Akan*[2]

ABSTRACT

In this chapter, we focus on routing in cognitive vehicular networks. Compared to other wireless networks, vehicular networks have a considerably more dynamic nature due to the speed of the vehicles. Connectivity changes rapidly as vehicles move around, as well as the channel properties due to the changes in the landscape. For the cognitive vehicular networks, another consideration is the rapid change in channel availability; as vehicles move, they go in and out of transmission range of primary users. Therefore, routing in cognitive vehicular networks is a very challenging task. We investigate conventional routing schemes in mobile ad hoc networks and also vehicular networks. We point out how they are not directly usable for cognitive vehicular networks and point out possible extensions to these schemes. We also lay out exploitable features of cognitive vehicular networks and indicate what can be done to take advantage of these feature. We investigate routing in CVNs according to various criteria such as communication paradigm, and routing approach.

1. Introduction

With the recent interest in providing ubiquitous wireless access, vehicular networks have attracted significant interest from the academic community as well as major automobile manufacturers and government bodies [1-3]. Considering the large number of vehicles on the roads and the ever-increasing demand for the wireless spectrum, it

---

[1] Research & Teaching Assistant, Next-Generation Wireless Communication Laboratory, Koc University.
[2] Faculty member, Director of Next-Generation Wireless Communication Laboratory, Koc University.
* Corresponding author: ozergul@ku.edu.tr

is envisioned that in the near future vehicles will have cognitive radio capability to combat the spectrum scarcity problem. The resulting network paradigm called cognitive vehicular network (CVN), has a very dynamic nature due to the high speed of the vehicles which causes channel conditions and spectrum availability to change rapidly.

One of the biggest challenges in such a dynamic network is routing. The initial research on routing for CVNs was mostly extensions to routing schemes proposed for vehicular ad hoc networks (VANETs), which were themselves modifications to routing schemes designed for mobile ad hoc networks (MANETs). However, there are some significant differences between the two in terms of routing as detailed in the next section.

2. Challenges and Exploitable Features of CVN

Routing in CVNs has some specific challenges. These mostly arise from the high speed of vehicles and the specific nature of traffic in CVNs. These challenges can be listed as follows:

- *Channel conditions changes*: These can be divided into two:
  - ○ *Environmental changes*: E.g., shadowing changes due to various buildings while a vehicle is moving in urban areas.
  - ○ *Fast fading*: High mobility causes rapid changes in received multipath components of a signal.
- *Topology changes*: Vehicles rapidly getting in and moving out of transmission range of each other cause fast changes in neighbor list of a node.
- *Node density changes*: Vehicles can go through dense highways, and then move to uncrowded streets. Therefore, total demand for bandwidth and network congestion may change drastically.
- *Spectrum availability changes*: Spectrum availability is location dependent. Vehicles may move from crowded commercial regions, where there are lots of primary users (PUs), to residential areas with few PUs and abundant available spectrum. Thus, bandwidth availability may be highly erratic.
- *Wide range of quality of service (QoS) needs*: QoS requirements change from high reliability and minimum delay for vehicle collision avoidance applications to high bandwidth for in-car multimedia infotainment applications. Reliability is vital for certain applications such as safety messages pertaining to vehicle proximity, road hazards, and other emergencies. Compared to conventional networks, the stakes are much higher since human life is at risk. Therefore, strict adherence to QoS requirements is of utmost importance.

While these challenges aggravate the implementation of effective routing mechanisms for CVNs, there are also some features specific to CVNs that can be exploited.

- *Abundant energy*: Since the transceiver is a part of the vehicle, energy consumption is not as much of a concern as it is for conventional hand held devices.

- *No size and weight concern*: The radio-equipment is part of the vehicle and will not be carried around like handheld wireless devices. Therefore, size and weight are less of a concern.

- *Higher cost possible*: Since the cognitive radio (CR) unit is part of a much bigger cost, i.e., that of the vehicle, adding a small increase to the cost of CR unit by adding more capabilities is viable.

As a result following abilities can be added to CVNs:

- *More processing power and memory*: CVNs can use more sophisticated spectrum sensing, spectrum management and cooperation schemes compared to conventional cognitive radio networks (CRNs).

- *Multiple radios:* Solutions that involve multiple transceiver units are viable. E.g., one for common control channel (CCC) monitoring, one for spectrum sensing and another for data transmission.

- *Higher transmission power:* Transmission range can be increased or concurrent transmission from different transceivers can be made where appropriate.

- *Predictable mobility patterns:* In VN, moving patterns are more predictable since vehicles have to move on roads. By obtaining data on the layout of roads, more accurate mobility prediction algorithms may be utilized.

- *Precise location information:* Future vehicles will be equipped with Global Positioning System (GPS) devices, which introduce features such as:
  - Location information is available at all times.
  - Synchronization is easier due to very accurate GPS clock.

In the following sections, we investigate the existing routing schemes for CVNs, as well as some earlier solutions developed for MANETs and VANETS that have been inspirations for the mentioned CVN routing solutions. We conduct our investigation according to various categorizations as shown in Figure 1 and provide an extensive review of the literature.

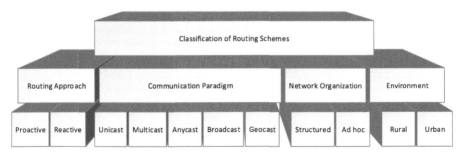

Figure 1. Classification of routing schemes.

3. Approaches to Routing in CVN

There are two main routing approaches. Proactive routing schemes [22, 23, 24, 27] update their routing tables with periodic messages regardless of whether there is a routing request or not. Reactive routing schemes [25–27] try to find routes when a routing request arrives.

3.1 Proactive

The advantage of proactive routing is that packets can be routed immediately since the routing tables are up-to-date all the time. The disadvantage is that since CVNs are highly dynamic, the frequency of route update messages must be considerably higher than conventional MANETs. Since these messages are sent regardless of network load or available bandwidth, the overhead may be excessive.

As the number of nodes gets higher, routing tables become harder to maintain, especially if the networks has a highly varying topology as in CVNs. This scaling problem makes pure use of proactive schemes unfit for CVNs, since they ultimately aim to connect a vast number of vehicles over huge geographical distances.

On the other hand, if the number of nodes is not high, optimal routing decisions may be made since all of the required information about links is maintained in the routing tables [16]. This suggests a hierarchical routing model, where proactive mechanisms are used in a higher layer only to connect predetermined special nodes (e.g., RSUs) to send packets to a certain region. Once the packet reaches the related region, reactive methods may be utilized. This idea was used in [27] for CVNs.

In terms of spectrum availability, in proactive schemes nodes perform spectrum sensing periodically and maintain a spectrum availability map. Most of the studies that determine routes by optimizing a certain metric inherently assume a proactive scheme [15, 17, 19, 28]. There are also proposal to modify routing protocols developed for MANETs for CVNs [24].

3.2 Reactive

Reactive routing schemes do not have the high overhead problem of proactive schemes. However, they are slower to respond to route requests, since the route is determined after the request by probing the nodes generally by flooding or similar types of messaging.

Reactive schemes are mostly used in ad hoc networks, where forwarding decisions are made by individual nodes as the packet moves towards the destination. Spectrum sensing is performed just before packet forwarding. This introduces certain delay to routing, and the most optimal route may not be chosen since global information is not available. However, the spectrum availability information and neighbor node list is current and the message overhead is kept to a minimum.

Generally, local optimization is made to minimize or maximize a certain metric at each hop instead of a global optimization [20]. Most used metrics are Expected Transmission Time (ETT), Expected Transmission Count (ETX), delay [18] and link stability [5]. In CVNs, network topology changes rapidly both due to high speed of the vehicles and due to changes in spectrum availability. Choosing next hops by trying to maximize may lead to the choice of channels which are used very frequently by the PU. Even if the data rate at that channel is high, PU arrival is also more likely. Upon PU arrival, a lot of time will be wasted on finding another available channel and negotiating spectrum assignment, which reduces the overall throughput. Therefore, link stability is a concern that is specific to VNs and more so to CVNs.

4. Communication Paradigm

Communication paradigm may be unicast, multicast, anycast, or broadcast. Unicast routing is from a single source to a single destination. It is generally used in vehicle infotainment and Internet access. Most of the research is on unicast routing [5, 18, 19, 24]. Multicast routing originates from a single source to a number of destinations. This is generally used when an application need to send messages to multiple users [29–31]. In anycast routing, the aim is to reach any one of the nodes in a certain group [32]. The difference between multicast and anycast is that in multicast the message is routed to all of the nodes in the group, while in anycast it is sufficient if only one of the nodes in the group receive the message. Broadcast is used to send messages to all of the nodes in the vicinity.

For VNs, there is another type of routing, called geocast. Geocast is actually a special form of multicast or anycast, where the destination group is identified by their location. Geocast is used to send warning messages to vehicles in a certain area (e.g., in case of an accident) or to gather information from them [33–35]. The communication paradigms are shown in Figure 2.

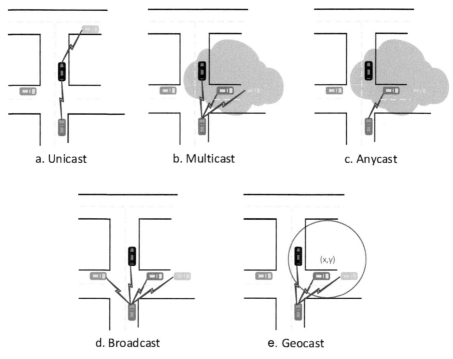

Figure 2. Communication paradigms.

Multicast, anycast, broadcast and geocast are particularly hard for CVNs, since common spectrum availability is harder to find for a group of nodes instead of a pair of nodes. To the best of our knowledge, there is still no research on any of these routing paradigms for CVN.

5. Network Organization

Routing in CVNs may be broadly divided into two based on network organization: Structured and ad hoc. Structured CVNs usually have road side units (RSUs) to control the communication. Routing schemes developed for structured CVNs generally employ a centralized approach where each RSU controls a certain area, collects information and routing demands from vehicles and disseminate channel allocation and routing decisions. On the other hand, in ad hoc vehicular networks there are no predetermined elements in the network structure. The whole network generally only consists of vehicles and there are no infrastructural elements to help with routing. These two possible network organizations are shown in Figure 3.

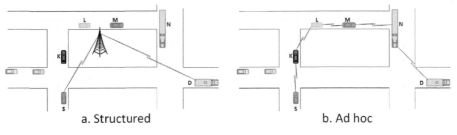

a. Structured   b. Ad hoc

Figure 3. Network organization.

5.1 Structured

Structured CVNs are formed using RSUs. Generally RSUs maintain or have access to spectrum availability map as well as location information of all vehicles inside their domain. In a sense they are the base stations (BSs) of the CVN. They either directly determine routing decisions [15], or help vehicles by answering their queries [16, 17].

An alternative method of using cognitive radio is to find active networks and utilize them. In [18] a cognitive communication device for CVNs is proposed. The device is capable of using various types of networks such as satellite, GSM, Wi-Fi through its multiple interfaces. A structured network architecture with RSUs and vehicles using this device is envisioned in which RSUs manage their respective sites the whole network is coordinated by a Services Access and Management Server (SAMS). Nodes that have access to networks with limited coverage (UHF, VHF, 802.11) run modified versions of well-known proactive routing protocol OLSR, called GPS enabled Optimized Linked State Routing Protocol for Multiple Interfaces (GOLSR-MI) while nodes that have access to global networks such as General Packet Radio Service (GPRS), Wireless Local Loop (WLL) and satellite uses the modified version of reactive routing protocol Ad hoc On-Demand Distance Vector (AODV), called AODV for Multiple Interfaces (AODV-MI).

5.2 Ad hoc

Ad hoc vehicular networks generally only consist of vehicles. Forwarding decisions are made individually at each hop. Most of the proposed solutions are geo-location based and assume that the locations of the destination and the neighbor vehicles are known. Since there is no coordinating entity, generally multiple radios per vehicle are assumed [5, 18, 19]. One of the radios is used for coordination between the neighbor vehicles, while other(s) are used for data exchange.

Routing schemes for ad hoc network structures cannot yield optimal solutions since global optimization is not possible. On the other hand, due to highly dynamic nature of CVNs the large control message overhead that is inevitable in centralized solutions is avoided.

The relay nodes can be chosen according to a number of considerations such as delay [18], path stability [5, 19] and expected transmission count (ETX) which is the average number of transmissions required to deliver a packet across a link [20].

6. Routing Based on the Environment

Solution approach may change considerably depending on the conditions of the area vehicles are in. There is a huge difference in the amount of available bandwidth between rural and urban areas. The traffic density may be higher than urban areas in highways, but the spectrum availability may be similar. Therefore, solutions developed assuming one type of environment may not be applicable to another. Even though research is more focused on routing in urban environment, there are also a few studies that target urban areas and highways for CVNs.

6.1 Rural Areas and Highways

Rural areas are characterized by lower vehicle density, higher spectrum availability, lower number of obstructions (i.e., lack of high buildings) and generally lower vehicle speeds. Therefore, it is generally more suitable to adopt routing schemes form MANET. However, in VNs in rural areas may be partitioned due to frequent losses of line-of-sight (LOS) to neighbor vehicles because of the curved roads and absence or low number of RSUs. Thus, in addition to spectrum awareness, specific modifications to adopted routing schemes are needed.

Greedy Perimeter Stateless Routing (GPSR) [4], and the routing scheme proposed in [5] were designed to address the issues of open environments for traditional VNs. GPSR works in two stages. This two-step approach is adopted by many of the later routing schemes, including those designed for CVNs. In the first stage, a greedy algorithm forwards packets to the nodes nearest to the target. As the packets get to the nearest connected node but it does not have direct connection to the destination stage two is operation is executed. In this stage, perimeter forwarding is used and packets are forwarded according to right hand rule to select the next route. This last stage is called the "recovery stage". We present detailed explanation when discussing types of routing in Section 7. There are also routing schemes that are more suitable for highways such as dynamic gradient-based routing (DGR) [6].

None of these schemes have spectrum awareness. However, they are solid approaches that can form the basis for cognitive routing for VNs in rural environments. One of the rare solutions for routing in rural CVNs is presented in [5], where routes are established by trying to maximize the expected duration of the link. Since both spectrum availability and network topology are highly dynamic, considering the lifetime of links in the route may considerably improve the communication performance.

6.2 Urban

Most of the research in routing for VNs is concentrated on urban environments. The main difference compared to routing in rural areas is the consideration of vehicle traffic on alternative roads. Some of the solutions find paths through fixed "anchor points" in the topology that help with routing [7, 8], while others discuss the advantages of dynamic selection of these anchor points [9, 10]. Some make use of static RSUs [11] and others argue busses and/or taxis operating in the city as mobile RSUs increases performance [12, 13].

One of the possible methods to identify spectrum availability in urban areas where shadowing degrades spectrum sensing performance is to use spectrum availability maps. These maps are either issued by an authority or formed cooperatively by gathering the spectrum sensing results of nodes in the locality. Using this idea, spectrum-map-empowered opportunistic routing (SMOR) was proposed in [21]. Authors introduce two algorithms. One for regular size networks and one for large scale networks where only local spectrum availability information is not sufficient to determine end-to-end routes. However, these algorithms do not address the problems caused by high speeds of vehicles in VNs.

Another problem specific to urban areas is related to GPS. Most of the routing protocols for VNs use geo-location based routing. Furthermore, spectrum maps also use location data, mostly obtained from GPS. Therefore, performance of most routing methods depends on the accuracy of GPS data. However, GPS connectivity may be problematic in streets surrounded by high buildings. These locations are called urban canyons. In [14], authors report that GPS signal strength may fall way below minimum in such scenarios. Even if an accurate spectrum availability map exists, routing will be adversely affected if location information is faulty or non-existent. Further research is needed in this area.

7. Types of Routing

Communication in CVNs is generally multi-hop. Routing schemes can be classified according to the methods they employ in choosing their next hop. The employed methods can be greedy routing, delay tolerant routing, cluster-based routing or QoS-based routing. Below we investigate each of these methods.

7.1 Greedy Routing

The most common routing method is greedy routing, where the next hop is chosen to be the node that is closest to the destination [4, 24]. This reduces the number of hopes and thus, theoretically the overall delay. Generally, the availability of GPS is assumed. In such cases, since the coordinates of both the destination and the neighbor nodes are known, the neighbor closest to the destination node can be determined. However, it is possible to have scenarios where none of the neighbors are closer to the destination than the current forwarding nodes. This is depicted in Figure 4. Node S wants to forward packets to the destination node D. However, none of its neighbors are located closer to D. By purely greedy forwarding, it is not possible to route the packets to the destination. However, there is a path through vehicles K, L and M to the destination. In such cases, where greedy algorithms fail certain recovery algorithms are proposed. The simplest recovery mechanism is flooding. However, while it is effective, it is highly inefficient. In [4], face routing is used for recovery. In face routing the so-called right-hand ruled is invoked, where if a closer node to the destination cannot be found, the algorithm walks around the faces of a planar graph, which are progressively closer to the destination vertex. This enables packets to be routed through vehicles K, L, M and N to D in Figure 4. A survey on recovery methods can be found in [36].

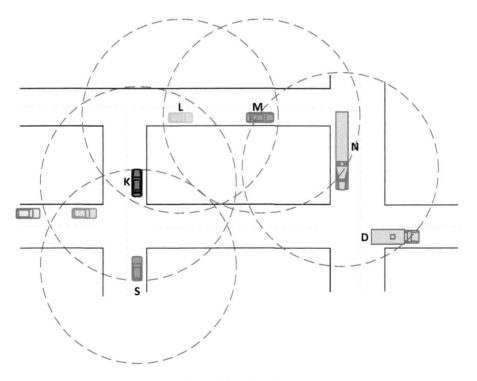

Figure 4. Greedy routing.

7.2 Cluster-based Routing

Aside from the need for recovery mechanisms, another problem of greedy algorithms is that in choosing the next hop closest to destination, they generally opt for the furthest node in their communication range. With high vehicle speeds, it is possible that these next hop vehicles go out of transmission range before the communication ends. Furthermore, due to the high distance, the channel between these nodes is not likely to have the highest data rate.

The routing takes place in two phases. In the first phase, cluster members send their packets to the cluster head. This is called inter-cluster communication. In the second phase, cluster head forwards the packets to other related cluster heads [28, 41]. Cluster head selection algorithms take into consideration parameters such as vehicle speed, number of neighbors, distance between neighbors, etc. In CVNs spectrum availability should also be taken into account. Nodes that have more available channels common with their neighbors are better candidates for cluster heads. On the other hand, unlike in MANETs, energy state of the nodes is not an issue in VNs.

Cluster-based routing is especially useful while disseminating security related information to vehicles within certain vicinity. In [42], vehicles are organized in zones, called peer spaces, according to their common interests. The number of peers inside a peer space is limited since there is no advantage for a peer to maintain knowledge for many others. This helps with reducing the overhead inside a peer space.

Cluster-based formations are also suitable for cooperative spectrum sensing. While cooperation increases spectrum sensing performance, the gain reduces as the number of cooperating nodes increase. The delay incurred by exchanging spectrum sensing data and negotiating among the cooperating nodes also increase with the number of nodes. Therefore, cluster-based methods, with limited number of nodes in each cluster tend to yield good results. Furthermore, coordination of spectrum sensing and routing is needed to obtain "silent" channel on which spectrum can reliable be sensed. Cluster-based mechanisms also help with such coordination. However, the amount of research on this area is limited [28, 41].

7.3 Delay Tolerant Routing (DTR)

In VNs network partitioning is a common occurrence since links may frequently be broken due to high vehicle velocity. Delay tolerant routing or opportunistic routing is used in such cases, where a node passes the packets to another node that travels in the direction of the destination. The intermediate node stores and carries the packets as it moves towards the destination and relays the packets once it gets into the communication range of the destination or the partition of the network that is connected to the destination.

In DTR, information is generally transmitted in as bundles of multiple packets. These "bundles" contain all relevant information such as protocol data, authentication data, etc., that is required for the completion of a transaction in one go [37, 38]. This is helps avoiding several round trips to exchange protocol messages and/or authentication.

DTR can be used both in ad hoc and structured networks. In ad hoc networks vehicles are the only means to "carry" the messages. In structured networks, RSUs may be used to store the bundle until a vehicle moving towards the destination arrives.

Since the network is not fully connected when a DTR scheme is used, delivery of the bundle is not guaranteed. Therefore, some schemes used multiple copies of the bundle [39].

A survey of routing protocols for vehicular delay tolerant networks can be found in [40]. However none of the investigated solutions are for CVNs. DTR in CVN is an open research area.

7.4 QoS-based Routing

CVN applications have a wide range of quality of service (QoS) requirements. Safety related applications such as vehicle collision avoidance have very low, strict minimum delay and reliability requirements, while in-car multimedia infotainment applications have high bandwidth requirements. Since human lives are at stake, strict adherence to QoS requirements is of utmost importance.

Link stability is one of the most important QoS factors. Aside from the essential reliability issues, it affects both data transmission rate and delay incurred by routing [5]. The scheme proposed in [43] uses vehicle position, speed and trajectory information to estimate quality factors of a route. These factors are used to form a metric called Expected Disconnection Degree (EDD) which is an estimation of probability that determines the breakability of route during given time period.

In [44], a cross-layer routing scheme is proposed, where path selection and routing decision are made according to different QoS requirements. Analysis of end-to-end reliability and total power consumption are provided. The multi-interface cognitive device proposed in [27] takes QoS requirements into consideration to prioritize the interfaces accordingly, though the exact mechanism is not explained in detail.

8. Security Issues

Malicious attacks related to routing either take advantage of the algorithmic properties of the routing protocols or advertise fake locations to their neighbor nodes. CVNs are especially challenging for secure routing algorithm design. One of the reasons for this is that due to mobility, an attacker may move into coverage of its victim, change its routing table with false advertisements and then move away before detection. In an environment where, legitimate nodes also move in and out of coverage range very frequently, such malicious information may be hard to detect [45].

With regard to geo-location based routing schemes, a malicious node may advertise fake locations for both itself and for other nodes. These attacks are difficult to mitigate. These may be countered to a degree with the predictability of vehicle movement in VNs.

VNs, especially in rural areas, have a single dimensional structure, i.e., the road that the vehicles travel. As the distance between nodes in a single dimensional structure increase, the probability of end-to-end connectivity decreases [46]. This makes an attacker easier to partition the network.

On the other hand, attacks that aim to deplete node battery are not effective in VNs. Also, nodes in VNs have higher memory and computation power and therefore, they are capable of supporting complex authentication mechanisms.

Acknowledgement

This work was supported by the Turkish Scientific and Technical Research Council (TUBITAK) under grant #110E249.

References

[1] Video Friday: Bosch and Cars, ROVs and Whales, and Kuka Arms and Chainsaws. IEEE Spectrum. 25 January 2013. Retrieved 7 March 2015.

[2] S.P. Wood, J. Chang, T. Healy and J. Wood. The potential regulatory challenges of increasingly autonomous motor vehicles. 52nd Santa Clara Law Review 4(9): 1423–1502.

[3] J.A. Ruiz, J.J. Gil, J.E. Naranjo, J.I. Suárez and B. Vinagre. Cooperative maneuver study between autonomous cars: Overtaking. Proc. 11th International Conference on Computer Aided Systems Theory EUROCAST 2007, pp. 1151–1158, February 2007.

[4] B. Karp and H.T. Kung. GPSR: Greedy perimeter stateless routing for wireless networks. Proc. 6th Annual International Conference on Mobile Computing and Networking, IEEE/ACM MobiCom 2000, pp. 243–254, 2000.

[5] J. Liu, P. Ren, S. Xue, H. Chen. Expected Path Duration Maximized Routing Algorithm in CR-VANETs, in Proc. 1st IEEE International Conference on Communications in China (ICCC), pp. 659,663, 15–17 Aug. 2012.

[6] Y. Guo, Z.-Z. Xu, C.-L. Chen and X.-P. Guan. DGR: Dynamic gradient-based routing protocol For unbalanced and persistent data transmission in wireless sensor and actor networks. Journal of Zhejiang University SCIENCE C, 12(4): 273–279, April 2011.

[7] C. Lochert, H. Hartenstein, J. Tian, H. Fussler, D. Hermann and M. Mauve. A Routing strategy for vehicular ad hoc networks in city environments. Proc. IEEE Intelligent Vehicles Symposium, pp. 156,161, June 2003.

[8] B.C. Seet, G. Liu, B.-S. Lee, C.-H. Foh, K.-J. Wong and K.-K. Lee. A-STAR: a mobile ad hoc routing strategy for metropolis vehicular communications. Proc. Third International IFIP-TC6 Networking Conference Athens, Greece, pp. 989–999, 2004.

[9] M. Jerbi, S.M. Senouci, R. Meraihi and Y.G. Doudane. An improved vehicular ad hoc routing protocol for city environments. Proc. IEEE International Conference on communications, pp. 3972–3979, 24–28 June 2007.

[10] S.M. Bilal, S. Mustafa and U. Saeed. Impact of directional density on GyTAR routing protocol for VANETs in city environments. Proc. IEEE 14th International Multitopic Conference (INMIC), pp. 296,300, 22–24 Dec. 2011.

[11] Y. Ding and L. Xiao. SADV: Static-Node-Assisted adaptive data dissemination in vehicular networks. IEEE Transactions on Vehicular Technology, 59(5): 2445–2455, Jun 2010.

[12] J. Luo, X. Gu, T. Zhao and W. Yan. A mobile infrastructure based VANET routing protocol in the urban environment. Proc. International Conference on Communications and Mobile Computing (CMC), 3: 432–437, 12–14 April 2010.

[13] H.-Y. Pan, R.-H. Jan, A.A.-K. Jeng, C. Chen and H.-R. Tseng. Mobile gateway routing for vehicular networks. Proc. 8th IEEE Asia Pacific wireless communication symposium (APWCS 2011), 2011.

[14] H. Kremo and O. Altintas. On detecting spectrum opportunities for cognitive vehicular networks in the TV white space. Journal of Signal Processing Systems, 73(3): 243–254, December 2013.

[15] C. Jiang, Y. Chen and K.J.R. Liu. Data-driven optimal throughput analysis for route selection in cognitive vehicular networks. IEEE Journal on Selected Areas in Communications, 32(11): 2149–2162, November 2014.

[16] M. Pan, P. Li and Y. Fang. Cooperative communication aware link scheduling for cognitive vehicular networks. IEEE Journal on Selected Areas in Communications, 30(4): 760,768, May 2012.

[17] C. Yang, Y. Fu, Y. Zhang, S. Xie and R. Yu. Energy-efficient hybrid spectrum access scheme in cognitive vehicular ad hoc networks. IEEE Communications Letters, 17(2): 329,332, February 2013.

[18] Y. Liu, L.X. Cai and X. Shen. Spectrum-aware opportunistic routing in multi-hop cognitive radio networks. IEEE Journal on Selected Areas in Communications, 30(10): 1958–1968, November 2012.

[19] X.X. Huang, D. Lu, P. Li and Y. Fang. Coolest Path: Spectrum mobility aware routing metrics in cognitive ad hoc networks. Proc. 31st International Conference on Distributed Computing Systems (ICDCS), pp. 182–191, June 2011.

[20] W. Kim, S.Y. Oh, M. Gerla and K.C. Lee. CoRoute: A new cognitive anypath vehicular routing protocol. Proc. 7th International Wireless Communications and Mobile Computing Conference (IWCMC), pp. 766–771, 4–8 July 2011.

[21] S.-C. Lin and K.-C. Chen. Spectrum-Map-Empowered opportunistic routing for cognitive radio ad hoc networks. IEEE Transactions on Vehicular Technology, 63(6): 2848–2861, July 2014.

[22] T. Clausen and P. Jacquet. Optimized Link State Routing Protocol (OLSR). RFC 3626, October 2003.

[23] C.E. Perkins and P. Bhagwat. Highly dynamic destination-sequenced distance-vector routing (DSDV) for mobile computers. Proc. Conference on Communications Architectures, Protocols and Applications, SIGCOMM 94, 1994.

[24] T. Stephan and K. Karuppanan. Cognitive inspired optimal routing of OLSR in VANET. Proc. International Conference on Recent Trends in Information Technology (ICRTIT), pp. 283–289, 25–27 July 2013.

[25] C. Perkins, E. Belding-Royer and S. Das. Ad hoc On-Demand Distance Vector (AODV) Routing. RFC 3561, July 2003.

[26] D. Johnson, Y. Hu and D. Maltz. The Dynamic Source Routing Protocol (DSR) for Mobile Ad Hoc Networks for IPv4. RFC 4728, February 2007.

[27] Z. Ahmed, H. Jamal, S. Khan, R. Mehboob and A. Ashraf. Cognitive communication device for vehicular networking. IEEE Transactions on Consumer Electronics, 55(2): 371–375, May 2009.

[28] X.-L. Huang, G. Wang, F. Hu and S. Kumar. Stability-Capacity-Adaptive routing for high-mobility multihop cognitive radio networks. IEEE Transactions on Vehicular Technology, 60(6): 2714–2729, July 2011.

[29] R. Jiang, Y. Zhu, X. Wang and L.M. Ni. TMC: Exploiting trajectories for multicast in sparse vehicular networks. IEEE Transactions on Parallel and Distributed Systems, 26(1): 262–271, Jan. 2015.

[30] H.S. Aghdasi, N. Torabi, A. Rahmanzadeh, M. Aminiazar and M. Abbaspour. Usefulness of multicast routing protocols for vehicular ad-hoc networks. Proc. Sixth International Symposium on Telecommunications (IST), pp. 459–463, Nov. 2012.

[31] A. Sebastian, M. Tang, Y. Feng, M. Looi. Context-Aware Multicast Protocol for Emergency Message Dissemination in Vehicular Networks. International Journal of Vehicular Technology, vol. 2012, Article ID 905396, doi:10.1155/2012/905396.

[32] L. Khan, N. Ayub and A. Saeed. Anycast based routing in vehicular ad hoc networks (VANETS) using Vanetmobisim. World Applied Sciences Journal, 7(11): 1341–1352, 2009.

[33] A. Bachir and A. Benslimane. A multicast protocol in ad hoc networks inter-vehicle geocast. Proc. 57th IEEE Semiannual Vehicular Technology Conference, VTC 2003-Spring, 4: 2456–2460, April 2003.

[34] Y. Li and W. Wang. Horizon on the move: geocast in intermittently connected vehicular ad hoc networks. Proc. IEEE INFOCOM, pp. 2553–2561, April 2013.

[35] A. Festag, P. Papadimitratos and T. Tielert. Design and performance of secure geocast for vehicular communication. IEEE Transactions on Vehicular Technology, 59(5): 2456–2471, Jun 2010.

[36] D. Chen and P. Varshney. A survey of void handling techniques for geographic routing in wireless networks. IEEE Communications Surveys and Tutorials, 9(1): 50–67, 2007.

[37] V.N.G.J. Soares, F. Farahmand and J. Rodrigues. A layered architecture for vehicular delay-tolerant networks. Proc. IEEE Symposium on Computers and Communications, pp. 122–127, 2009.

[38] V.N.G.J. Soares, J.J.P.C. Rodrigues and F. Farahmand. Performance assessment of a geographic routing protocol for vehicular delay-tolerant networks. Proc. IEEE Wireless Communications and Networking Conference, pp. 2526–2531, 2012.

[39] A. Casaca, J.J.P.C. Rodrigues, V.N.G.J. Soares and J. Triay. From delay-tolerant networks to vehicular delay-tolerant networks. IEEE Commun. Surv. Tutorials, 14(4): 1166–1182, 2012.

[40] N. Benamara, K.D. Singhb, M. Benamara, D.E. Ouadghiria and J.-M. Bonninb. Routing protocols in vehicular delay tolerant networks: A comprehensive survey. Computer Communications, 48: 141–158, July 2014.

[41] D. Niyato, E. Hossain and P. Wang. Optimal channel access management with QoS support for cognitive vehicular networks. IEEE Transactions on Mobile Computing, 10(4): 573–591, April 2011.

[42] I. Chisalita and N. Shahmehri. A peer-to-peer approach to vehicular communication for the support of traffic safety applications. Proc. 5th IEEE International Conference on Intelligent Transportation Systems, pp. 336–341, September 2002.

[43] Z. Mo, H. Zhu, K. Makki and N. Pissinou. MURU: A multi-hop routing protocol for urban vehicular ad hoc networks. Proc. IEEE Annual International Conference on Mobile and Ubiquitous Systems Workshops, pp. 1–8, July 2006.

[44] Z. Ding and K.K. Leung. Cross-layer routing using cooperative transmission in vehicular ad hoc networks. IEEE J. Sel. Areas in Commun., 29(3): 571–581, March 2011.

[45] I. Broustis and M. Faloutsos. Routing in Vehicular Networks: Feasibility, modeling, and security. International Journal of Vehicular Technology; Q4 2008, vol. 2008, p1.

[46] O. Dousse, P. Thiran and M. Hasler. Connectivity in ad hoc and hybrid networks. Proc. Joint Conference of the IEEE Computer and Communications Societies (INFOCOM '02) 2: 1079–1088, June 2002.

7

Connectivity Analysis and Modeling in Cognitive Vehicular Networks

Elif Bozkaya and *Berk Canberk**

ABSTRACT

This chapter presents an overview of the network connectivity problem in Cognitive Vehicular Networks (CVNs). Recent studies in CVNs have shown that dramatic changes in spatial and temporal topological behaviors of such wireless networks occur due to the unexpected mobility of vehicles, the increasing use of vehicular applications, limited transmission ranges of Road Side Units (RSUs) and channel utilization ratios of primary users. Such a highly fluctuating network triggers possible interference with primary users by resulting in degradation of CVNs' performance and this brings a crucial problem in terms of maintaining network connectivity. Establishing continuous communication and disseminating online information result with intermittent connectivity. All these challenges cause degradations both in the vehicle satisfaction and wireless communication quality in CVNs. Therefore, in this chapter, we concentrate on the network connectivity problem and present a connectivity modeling in CVNs.

Wireless communication systems explore innovative approaches and opportunities available in business and society through technology. Within the last 10 to 15 years, wireless communication networks got popular and common with the help of the idea of "Internet of Things" that all physical object could eventually communicate data over the internet to other connected devices, from automobiles to household devices.

Department of Computer Engineering, Istanbul Technical University, Istanbul, Turkey.
  Email: bozkayae@itu.edu.tr
* Corresponding author: canberk@itu.edu.tr

These developments towards wireless networks have considerably attracted in automotive industries. One of the most important reason is that every year, traffic accidents cause to heavy casualties due to the unawareness of drivers, misjudgment or operation miss. In order to avoid traffic accidents, and improve traffic safety and efficiency, Vehicular Networks, also known as Vehicular Ad hoc NETworks (VANETs), have been a promising technology by offering many advantages not only safety issues but also entertainment issues. However, how can be vehicles connected and communicate with each other or infrastructure? More importantly, what are the advantages of all that connectivity in vehicular networks? AT&T's chief executive Glenn Lurie expressed that "better voice activation, better voice diagnostics-all the things you want so your hands stay on the wheel and your eyes stay on road".

Although the U.S. Federal Communications Commission (FCC) has allocated 75 MHz in the 5.9 GHz frequency band for Dedicated Short Range Communication (DSRC) with the aim of usage in vehicular networks, increasing bandwidth requirement in vehicular applications causes a crucial problem; *spectrum scarcity*. This problem has been addressed by a new communication paradigm, called Cognitive Radio (CR).

Cognitive Radio enables to utilize vacant licensed spectrum bands by unlicensed users and improves spectrum utilization for unlicensed users so that inefficient usage of existing spectrum can be prevented. CR technology uses Dynamic Spectrum Access (DSA) techniques to achieve an effective spectrum management. As defined in [1]:

CR techniques provide the capability to use or share the spectrum in an opportunistic manner. DSA techniques allow the cognitive radio to operate in the best available channel.

An efficient mechanism enables to utilize spectrum resources with a continuous and robust connectivity among vehicles as unlicensed users. This mechanism includes the following functions [1]:

- *Spectrum sensing:* Detecting spectrum opportunities in order to avoid the interference with primary users.
- *Spectrum decision:* Deciding best available channels to satisfy QoS requirement of vehicles based on vehicular applications.
- *Spectrum sharing:* Providing a fair share of available spectrum resources among vehicles.
- *Spectrum mobility:* Switching channel due to the existence of primary users or better channel conditions to maintain continuous network connectivity.

Hence, CR technology is also a potential approach for an effective spectrum management for vehicles as unlicensed users in order to support growing demand in vehicular networks. Therefore, in Cognitive Vehicular Networks (CVNs), vehicles as unlicensed users can have an opportunity to utilize available licensed spectrum band as efficiently as possible. However, when compared with Cognitive Radio Networks (CRNs), high fluctuations in the available spectrum band occurs depending on not only channel usage pattern of primary users but also relative motion between vehicles. Therefore vehicles cannot continuously obtain a channel throughout all communication periods and this brings a crucial problem in terms of maintaining network connectivity.

Establishing continuous communication and disseminating online information result with an intermittent connectivity.

In addition to channel usage status (idle or busy) by primary users, high mobility of vehicles, increasing use of vehicular applications, limited transmission range of roadside units (RSUs) and vehicles cause dramatic changes in spatial and temporal behaviors of the network topology. All these challenges cause degradations both in the vehicle satisfaction and wireless communication quality in CVNs by resulting with high number of channel switching. Therefore, continuous allocation of spectrum is more challenging in CVNs to maintain network connectivity.

Based on the connectivity maintenance among vehicles and between vehicles and RSUs, this chapter addresses the following contributions:

- To address the connectivity problem in CVNs, we firstly need to look into the network connectivity in CRNs.
- Then, the challenges of CVNs peculiar to connectivity are investigated.
- Depending on connectivity challenges, the details of an existing work related to connectivity modeling in CVNs are presented.

1. Network Connectivity in Cognitive Radio Networks

Cognitive Radio Network mainly focuses on spectrum decision, spectrum sensing, spectrum sharing, spectrum mobility functionalities [2] for an efficient spectrum management so that available spectrum resources can be used as efficiently as possible among secondary users and network performance can be improved. An efficient mechanism enables to utilize spectrum resources with a continuous and robust connectivity.

Connectivity means that all nodes within the transmission range can be connected and communicate with each other on the same channel and spectrum band for a successful transmission. In CVNs, connectivity can be achieved if and only if this channel and spectrum band are not used by a primary user and then a reliable communication is provided among nodes. In CVNs, network connectivity mainly depends on the transmission range of primary users and secondary users, distance among nodes, the number of available channels and activities of primary users. Especially, primary user activities affect the connectivity of secondary networks. Hence, the density of primary users, the probability of primary users' activities for each spectrum band should be analyzed in an efficient manner. Moreover, due to the existence of primary users with heterogeneous Quality of Service (QoS) requirements, each channel characterization changes spatially and temporally so that network connectivity problem will be more challenging when compared to traditional wireless ad hoc networks.

In literature, connectivity problem is analyzed and designed via the graph theory in many researches for CVNs. Thus, network topology is modeled by a graph. The graph is denoted as $G(V, E)$ and consists of a set of nodes V and a set of edges E, which shows the communication link between nodes. An edge between nodes (i, j) may exist if the distance between nodes i and j is less than or equal to transmission range.

This condition is represented by an adjacency matrix, which shows the connectivity between nodes.

When a network topology is represented by a graph, the connectivity becomes the smallest number of communication links whose breakdown would jeopardize the communication in the network topology [4]. Therefore, higher connectivity will result more reliable communication in the network. Here, we provide some definitions [4] and theorems [24] [33] [34] related to connectivity analysis and modeling in graph theory.

- *Definition 1*: A graph G is represented by $G(V,E)$, where V is the set of nodes and E is the set of edges. G is k-connected is $|G| < k$ and G-n is connected for every set $n \subseteq V$ with $|n| < k$. k is the greatest integer and G is the k-connected, denoted as connectivity $\kappa(G)$ of G.

- *Definition 2*: A vertex cut is a subset of the vertices of a graph, and removing them disconnects the graph. A vertex cut of G is a subset V' of V such that G-V' is disconnected and a vertex cut of k elements is k-vertex cut. A complete graph has no vertex cut. If G has at least one pair of distinct nonadjacent vertices, the connectivity $\kappa(G)$ of G is the smallest k for which G has a k-vertex cut. Otherwise, $\kappa(G) = V$-1 and if G is disconnected, then $\kappa(G) = 0$.

- *Definition 3*: An edge cut is a subset of the edges of a graph, and removing them disconnects the graph. A subset of E of the form $[S, \overline{S}]$ where S is a nonempty subset of V. An edge cut of k elements is k-edge cut. If G is 1-connected graph and E' is an edge cut of G, G-E' is disconnected so that the edge connectivity $\kappa'(G)$ of G is smallest k for which G has a k-edge cut. If G is disconnected or V = 1, then $\kappa'(G) = 0$.

- *Theorem 1 (Menger's Theorem)*: The minimum number of vertices separating two nonadjacent vertices v_i and v_j is the maximum number of disjoint $v_i - v_j$ paths.

- *Theorem 2 (Whitney's Theorem)*: A graph G is n-connected if every pair of vertices of G are connected by at least n internally-disjoint paths.

These definitions and theorems are used in many connectivity analysis and modeling researches in CRNs. In this respect, network connectivity problem has been extensively studied and especially graph theory has been widely used in literature. In [22], the authors analyze the connectivity of CRNs from the perspective of probability and show the relationship among connectivity and density of primary users, density of secondary users, transmission radius of secondary users. [16] studies the impact of interference over the network connectivity in CRNs. By using percolation theory and stochastic geometry, it is observed that connectivity of CRNs will not adversely affect from interference. [12] investigates connectivity challenges in CRNs and compares with mobile ad hoc networks. A mathematical model is proposed in order to show the relation between connectivity and network parameters, i.e., secondary user's density, primary user's density, the number of channels, operating frequencies and the distribution of primary users on each channel. In [15], the authors propose a new connectivity metric, termed 'cognitive natural connectivity', under single and multi-primary user scenarios for CRNs. This metric has a similar performance with route availability metric which shows the probability of finding a route. Moreover, the complexity is significantly reduced. [11] analyzes the impact of primary users on the connectivity of secondary users in CRNs and show the relation of transmission range of primary users and secondary users on the network connectivity. [23] proposes a cognitive radio graph model which introduces survival probability by considering the number of channels and activities of primary users. Theories and techniques from continuum percolation are used to ensure dynamic connectivity in CRNs.

All these aforementioned researches being identified in CRNs, CVNs will show different characterization in terms of connectivity due to the highly dynamic vehicular network topology. Therefore, in the next section, we present the challenges of connectivity in CVNs.

2. Challenges of CVNs Peculiar to Connectivity

The following challenges peculiar to connectivity are mainly analyzed in this section. The details and descriptions will provide a better understanding of network connectivity analysis and modeling in CVNs.

- High vehicular mobility
- The increasing use of vehicular applications
- Spatial and temporal changes of available channels
- Limited transmission range of vehicles

2.1 High Vehicular Mobility

Mobility arises as one of the distinct characterization in CVNs. Depending on traffic density, informations of vehicles including position, speed, direction, changes over time and space, and these cause dramatic changes in vehicular network topology. As shown in Figure 1(a), vehicles can communicate with each other or RSUs within the transmission range of each other. After a time period, as seen in Figure 1(b), due to the movement of vehicles, available channels, QoS requirements and communication requests of current vehicles, the distribution of vehicles may vary by resulting with dynamic topology changes.

Although the movements of vehicular nodes are predictable, which are constrained to physical environment such as buildings, obstacles, trees, each geographic region shows different characteristics.

More specifically, spatial and temporal changes, relative speeds of vehicles, even when moving in the same direction, geographic conditions, changing radio propagation effects lead to attenuation of the signal and intermittent connectivity in CVNs. For example, Figure 2 shows the trade-off between link life and radio range depending on the direction of vehicles. Communication durations among vehicles traveling in opposite directions are very shortlived when compared to communication durations among vehicles traveling in the same direction. One of the important results in terms of vehicular mobility is that the extremely limited nature of links, even for vehicles traveling in the same direction with long radio ranges. Specifically, with a long transmission range of 500 ft, these links last approximately 1 min on average [3].

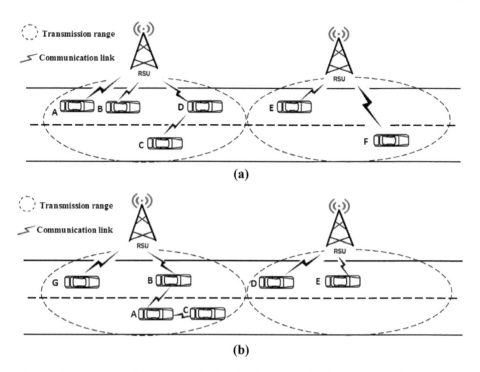

Figure 1. Changing network topology due to the vehicular mobility (a) Example of showing the vehicles within the transmission range of RSUs in the first situation (b) Example of showing as a result of movement of vehicles in the second situation.

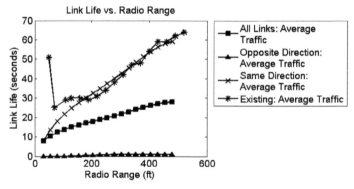

Figure 2. Communication durations w.r.t. radio ranges [3].

In addition, vehicular mobility strongly influences over spectrum management in terms of the correctness of the sensing information. For example, while low mobility enables to collect more samples per unit area, high mobility allows vehicles to move away from a shadow region as quickly as possible. Thus, spectrum management in CVNs may vary based on the characteristics of environment and the speed of vehicles. A vehicle can collect signal samples at the different locations. Many collected samples by all vehicles are utilized in the function of spectrum decision and it helps to reduce the risk of incorrect decision caused by shadowing effects [14].

Moreover, the selection of mobility model is of great importance in order to maintain a robust and continuous network connectivity. Vehicular ad hoc networks are separated from traditional mobile ad hoc networks (MANETs) with the property of high mobility. Therefore, commonly used mobility models in MANETs, such as Random Walk [36], Random Waypoint [37], are insufficient not to reflect a realistic vehicular environment due to the changes of node mobility depending on traffic density. In this respect, there exist many mobility models in literature for vehicular networks such as Manhattan mobility model, Freeway mobility model. However, these models also restrict vehicular mobility. For example, Freeway mobility model uses bi-directional multi-lane freeways and the movement of vehicles is restricted by lanes. Manhattan mobility model uses the grid road topologies and probabilistic approach in the choice of direction. Therefore, novel analytical mobility models are proposed [9] [25] in many researches.

In literature, mobility models are generally divided into two main categories [35]; *macroscopic description* involves restricted vehicle movements. To analyze the correlation between vehicular mobility and network connectivity, traffic stream models relate among average speed of vehicles, traffic density and traffic flow. *Microscopic description* models each vehicle behaviors as a distinct entity, but computationally more expensive [18].

2.2 The Increasing Use of Vehicular Applications

Due to the vehicular network technology which enables to obtain online and accurate information between vehicle to vehicle (V2V) and vehicle to infrastructure (V2I), the amount of safety applications has been increased. Thanks to safety applications, not only traffic safety and efficiency have been improved, but also driving comfort for drivers has been maintained such as monitoring, tracking, routing, solving congestion problem and situation awareness. Moreover, with recent advances in the access technologies, such as WiMAX, WiFi, 3G, infotainment applications have also emerged (e.g., video streaming, web browsing).

The consequences of these emerging applications also trigger to growing bandwidth demands by resulting with spectrum scarcity problem. Although CR technology is a potential solution to utilize unused spectrum band for vehicles as secondary users, available spectrum resources cannot be used throughout whole communication period due to the primary user activities and there is a strong need to maintain connectivity for these vehicular applications in CVNs.

2.3 Spatial and Temporal Changes of Available Channels

The utilization of each spectrum band changes spatially and temporally due to the existence of primary users and vehicles as secondary users in CVNs. Depending on primary user activities and the wireless communication quality, vehicles can dynamically need to switch current channel. Vehicles should adapt changing network environment by detecting best available channels and avoiding interference with primary users.

Dynamic Channel Selection

Channel switching is required when a primary user activity is detected or a better channel condition is needed. Therefore, vehicles need to exploit to channel status to check the availability of channels at all times. High fluctuation in the available spectrum band causes frequent channel switching by affecting the network performance and the quality of network connectivity adversely in CVNs. Each channel switching results with an intermittent connectivity and communication duration continuously changes depending on network topology. When the number of channel switching increases, the quality of network connectivity will be effected by causing communication disruptions. Therefore, the quality of communication is related to determine the best available channels dynamically.

In this respect, dynamic channel selection algorithms are proposed to determine the best available channels for data communication in many works. The number of channel switching as a performance parameter is also calculated to perform efficient spectrum management.

In [9], the authors propose a non-cooperative congestion game to exploit channel access opportunities. Spatial distribution and temporal channel usage of primary transmitters and mobility pattern are modeled. A distributed spectrum access algorithm is derived to access multiple channels from a game-theoretic perspective and the existence of the pure Nash equilibrium (NE) is also proved. Moreover, the proposed model is analyzed with uniform MAC and slotted ALOHA. In [25], the authors analyze the problem of optimal channel access by presenting a framework for cluster based communication in CVNs. Share-use channels and exclusive-use channels are considered to access the channels opportunistically and the reservation of a channel for dedicated access by vehicles. Moreover, cluster size control is adapted to maximize the utility of data transmission. [10] focuses on exploiting available channels and then proposes a cognitive channel hoping protocol to maintain network connectivity. Multiple channel selection and single channel selection are evaluated over the proposed protocol and multi-channel hopping improves network performance over single channel hopping with the proposed model. [28] proposes a channel selection algorithm and a distance-based multi-dimensional indexing approach in order to enable learning of vehicular environment in vehicular dynamic spectrum access networks (VDSANs). In this way, the experience about the location and requirements of vehicular applications are used to manage the resources efficiently. In [31], the authors propose a dynamic channel selection scheme in order to maximize data transmission within the period in multi-hop VANETs by using DSA techniques.

In summary, dynamic channel selection algorithms can be divided into two main categories in order to maintain full network connectivity:

- *Multi-channel selection:* Vehicles as secondary users can sense multiple channels in licensed bands, by using traditional sensing techniques in the literature of CR technology such as matched filter detection, energy detection, transmitter detection, etc. [1], in order to obtain the information of channel availability. Then vehicles can access one channel at a time when the channel is detected as available and switch to another suitable channel dynamically when channel is detected as unavailable.

• *Single-channel selection:* Vehicles can access only one channel throughout entire communication period depending on availability of channels.

TV White Spaces

CVNs are deployed over a wide range of the spectrum bands. One of the candidate of usable spectrum band is UHF television frequency, as often termed TV white spaces. FCC has allowed to vehicles to operate in the television spectrum when the spectrum is not being used by primary users and this band, which is between 470–698 MHz (Channel 14–Channel 51), is often analyzed in many researches based on CVNs due to the static channel usage pattern.

In this respect, in [7], reinforcement learning is used to achieve dynamic channel selection in a vehicular dynamic spectrum access (VDSA) environment. Number of channel switching, interference and throughput are analyzed by utilizing TV white spaces. [26] [27] describe the available TV channels in a specific geographic area. The authors present the implications of the noncontiguous channel availability in the TV spectrum on the design of a CR transceiver and devise a quantitative model based on spectrum measurements. One of the observation is that spectral occupancy changes various locations along the highway. However, this change is acceptable level so that it allows the vehicles to sense the frequency of TV broadcasts and change the frequency in order to avoid interruption or interference with TV signals.

Spectrum Coordination

Spectrum coordination is required to provide a fair share of available spectrum resources among vehicles. Therefore, an efficient spectrum access mechanism enables an efficient spectrum utilization and fair share of spectrum resources. An efficient spectrum management includes detecting spectrum opportunities, deciding best available channels to satisfy QoS requirement of vehicles based on vehicular applications and, using the channels by avoiding the interference with primary users in a fair manner.

In this respect, [21] applies Belief Propagation model for collaborative spectrum sensing in CVNs. In addition to local observation of vehicles, the informations received from other vehicles in the topology are utilized to obtain a stable information about the existence of primary users. Then, this spectrum sensing mechanism helps to vehicles by making decisions for appropriate spectrum bands. In [32], the authors propose a framework for spectrum sensing coordination in CR-VANETs. Each vehicle defines its spectrum sensing activities and, fine sensing activities of nearby vehicles are coordinated and scheduled in order to obtain an efficient sensing mechanism in CR-VANETs. [13] proposes a collaborative spectrum management framework in order to determine the accuracy of spectrum sensing, share spectrum information and detect spectrum opportunities at future locations. In [17], the authors analyze the channel allocation problem for multi-channel CVNs to maximize throughput for all vehicles. [29] presents a mechanism for connectivity management on CVNs which enables the use of channels according to the movement of vehicles and application requirements for reliable data delivery.

Moreover, queuing theory in terms of connectivity modeling has been extensively investigated in many researches to share available spectrum band among vehicles in a fair and efficient manner. Figure 3 generally represents a multi-channel queuing system. Here, channels are modeled as servers and communication requests of vehicles are evaluated according to availability of channels.

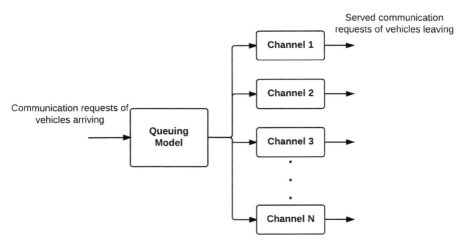

Figure 3. Multi-channel queuing system.

In queuing disciplines, the most common used approach is first-in first-out (FIFO). However, especially, when considered to safety applications, this is not the only possible approach. Messages should be categorized depending on types of vehicular applications. For example, a warning message in safety applications is higher priority than video streaming in infotainment applications. Moreover, in emergency situations, the message with highest priority is immediately transmitted even if a message with lower priority is already in transmission. Here, two priority discipline can be considered in CVNs: preemptive and non-preemptive. In preemptive case, while the message with highest priority is allowed to transmit immediately, the lower priority message is preempted and then retransmitted after message with highest priority is transmitted. The second situation, in non-preemptive case, the message with highest priority enters the head of the queue but this message does not serve until the ongoing transmission is completed.

In this respect, in [30], the authors examine the VDSA system by using queuing theory via multi-server multi-priority approaches and calculate transmission latency and the probability of all channels is busy over TV white space. [20] proposes M/M/1+G queuing model to analyze service requests of vehicles and characterizes the dynamics of the communication system under limited spectrum resources scenario in

DSA based vehicular communication. Average number of service requests, average waiting time, the probability of expiry as performance parameters are evaluated in the proposed queuing model. [8] presents a feasibility analysis based on queuing theory approach for VDSANs. Vacant TV channels are modeled as available servers and M/M/m and M/G/m queuing system with FIFO queue and priority queue approaches are evaluated in terms of the probability that all channels are busy and response time.

2.4 Limited Transmission Range of Vehicles

Network connectivity is required to establish communication and disseminate online information among vehicles and between vehicles and RSUs in CVNs. Limited transmission range causes short communication duration among vehicles and between vehicle and RSUs by resulting in intermittent connectivity and this brings many drawbacks such as degradation of wireless communication quality and frequent channel switching. Moreover, rapid changes in network topology result in short link lifetimes and while utilizing the spectrum opportunities, continuous connectivity cannot be achieved due to the movement of vehicles.

While safety is a major factor behind vehicular communications, vehicular networks provide a wide range of applications with different purposes. Especially, infotainment applications have become more popular due to the emerging applications with *mobile internet access*. All vehicular applications require information provided by other vehicles and RSUs. Therefore, transmission range of vehicles and RSUs have a major impact by affecting the communication duration in CVNs.

Vehicle-to-vehicle (V2V) communications allow short and medium range communications among vehicles as seen in Figure 4(a). In addition to limited range, due to the primary user activities, many critical V2V applications such as collision avoidance, monitoring, up-to-date traffic information, can be interrupted after a short communication period. Nevertheless, this drawback of V2V communications can be solved with the integration of RSUs. The deployment of RSUs enables to improve network performance by extending wireless coverage area in CVNs as seen in Figure 4(b). Here, RSU plays a significant role in the topology by gathering all global and local informations on traffic and road conditions such as position, speed, direction, acceleration information, channel availability status, and also RSU may propose some behaviors to vehicles within the geographical area of itself. Therefore, each vehicle registers to RSU when they come into the transmission range of it.

In literature, there exist many researches to show the trade-off between network connectivity and transmission range. In [19], the authors analyze the relationship among network connectivity, traffic density and transmission range and present a theoretical analysis of connectivity depending on size of transmission range. [6] analyzes the communication durations of vehicles within the transmission range of a RSU by proposing cognitive channel selection algorithms in CVNs. [3] shows relationship between network connectivity and transmission range between vehicles traveling in the same direction. It is showed in simulation environment that when transmission range of vehicles increases, network connectivity also increases.

In the next section, based on these aforementioned challenges, a connectivity modeling will be presented in CVNs.

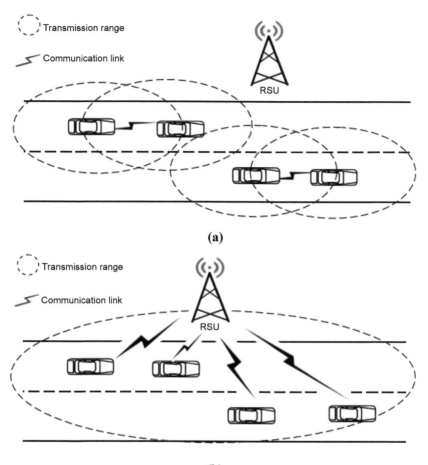

Figure 4. Connectivity analysis for two different CVN patterns (a) V2V communication (b) V2I communication.

3. Connectivity Modeling in CVNs

This section gives the details of an existing work as an example of connectivity modeling [5] based on the challenges of CVNs. In this research [5], to maintain robust and continuous connectivity, a centralized network topology is considered between RSU and vehicles. Here, channel activities and the communication requests of vehicles are modeled and then vehicle satisfaction ratio and channel switching as performance parameters are evaluated with the proposed dynamic channel selection algorithms. Figure 5 is proposed which is implemented into RSU and each module is modeled as follows:

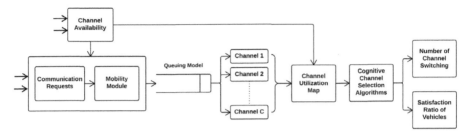

Figure 5. Connectivity model using queuing theory [5].

3.1 Channel Availability Module

In this module, each channel is modeled as idle or busy in each time slot which is represented with binary variables, and distributed according to Poisson process in time interval [1,t] as shown in the example Figure 6. Channel availability is defined as available or unavailable depending on spatial and temporal changes of channels. 1 and 0 values represent to busy and idle time intervals, respectively.

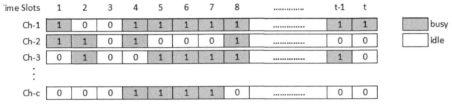

Figure 6. Example of showing for channel availability module.

3.2 Communication Requests Module

This module models the communication requests of vehicles within the transmission range of RSU. Communication requests are represented with binary variables, and distributed according to Poisson process in time interval [1,t] similar to channel availability module as shown in the example Figure 7.

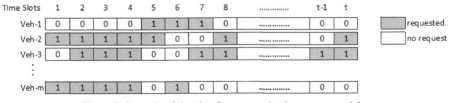

Figure 7. Example of showing for communication request module.

3.3 Mobility Module

This module analyzes the arrival and departure time of each vehicle within the transmission range of a RSU depending on the velocity of vehicles and traffic density. RSU has a limited transmission range and all vehicles have to register to RSU when they come into its range. Depending on traffic density, each vehicle moves at variable velocity, v', which are between $[v_{min} \leq v' \leq v_{max}]$, and RSU determines time period of each vehicle how long will remain in its transmission range (departure and arrival time). The movement of vehicles is considered normally distributed zero mean and unit variance.

3.4 Queuing Model

In this research, M/M/m queuing model with FIFO approach is considered to analyze the connectivity modeling. Channels are modeled as servers. Communication requests are evaluated in each time slot for each vehicle. The total number of communication requests in each time slot is compared with the number of available channels in the same time slot. If there is an available channel for each communication request to serve, all requests are accepted or otherwise requests until available channels are served and the remaining requests are waited in the queue until next time slot to check the availability of channels. Moreover, for further detail about the calculation of queuing model, [5] can be examined.

3.5 Channel Utilization Map Module

RSU updates all information of vehicles and channel availability status in each time slot. Depending on communication requests and channel availability, RSU obtains channel utilization map for each vehicle in each time slot as shown in Figure 8.

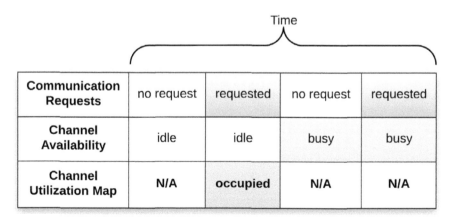

	Time			
Communication Requests	no request	requested	no request	requested
Channel Availability	idle	idle	busy	busy
Channel Utilization Map	N/A	occupied	N/A	N/A

Figure 8. Channel Utilization Map [5].

With the help of this channel utilization map, channel utilization ratios are calculated for each vehicle as follows:

$$U_{c_i}^m = \frac{\sum_{j=1}^{t} [U]_{c_i}^m}{t}, \quad \forall i \in [1,2,\dots c], \quad \forall j \in [1,2,\dots t] \tag{1}$$

where $U_{c_i}^m$ is utilization ratio of channel i for vehicle m, c is the number of channels, $c_i \in c$ and $[U]_{c_i}^m$ is the utilization matrix of ith channel for vehicle m.

After this channel utilization map and utilization ratios are obtained for each vehicle by RSU, each vehicle decides a cognitive channel selection algorithm (multi-channel or single channel selection) to maintain continuous network connectivity with minimal channel switching.

3.6 Cognitive Channel Selection Algorithms

In this module, six novel cognitive channel selection algorithms are determined. In order to analyze connectivity modeling, two performance parameters are defined and depending on these parameters, multi-channel and single channel selection approaches are evaluated.

- *Satisfaction ratio, ξ*: Total usage ratio of channels with respect to the proposed channel selection algorithms.
- *The number of channel switching, χ*: The total number of channel switching throughout the entire communication period.

Here, the main objective is to maximize satisfaction ratio while minimizing the number of channel switching by defining the best available channels with multi-channel or single channel approaches. With these motivations, pseudo codes of algorithms and explanations are as follows:

Alg. 1 is based on multiple channel selection by maintaining maximum connectivity with frequent channel switching. In Alg. 1, each vehicle determines the beginning indices of available and unavailable transmission periods in its utilization map as shown in example Figure 9. In Figure 9, 0 and 1 values represent available and

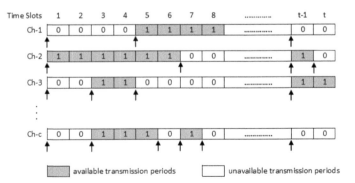

Figure 9. Determining beginning indices of consecutive available and unavailable periods in utilization map [5].

unavailable transmission periods, respectively. Each vehicle calculates consecutive available transmission intervals in its channel utilization map and unavailable intervals are eliminated. In each channel switching, a large amount of transmission interval is selected. Every selection is based on higher utilization in order to minimize channel switching. In this algorithm, vehicles dynamically switch to another suitable channel whenever a channel is detected as unavailable and all idle time slots are used to obtain maximum satisfaction ratio by resulting frequent channel switching.

Algorithm 1 Connectivity Maximization [5]

Input: $[U]_{ext}$

Output: ξ, χ

Temp. Variables: $\gamma, \Psi, C, C_{index}$

1:   $\xi, \chi \leftarrow 0$
2:   **for** i=1 to m **do**
3:     **for** j=1 to c **do**
4:       $\gamma \leftarrow$ Specify beginning index of consecutive available and unavailable transmission periods corresponding to $[U]_{ext}$
5:     **end for**
6:   **end for**
7:   **for** i=1 to m **do**
8:     **for** j=1 to c **do**
9:       Eliminate unavailable intervals in γ and $[U]_{ext}$
10:     **end for**
11:   **end for**
12:   **for** i=1 to m **do**
13:     **while** true **do**
14:       **for** j=1 to c **do**
15:         $\Psi \leftarrow$ Calculate consecutive available transmission periods according to γ and $[U]_{ext}$
16:         $C \leftarrow \max \{\Psi\}$
17:         $C_{index} \leftarrow$ beginning index of selected C
18:       **end for**
19:       **for** i=1 to m **do**
20:         **if** no vehicle is using the channel in the specified time slots **then**
21:           Wait until next slot
22:         **end if**
23:         $\xi \leftarrow \xi \cup \{C\}$
24:         $\chi \leftarrow \chi \cup \{C_{index}\}$
25:       **end for**
26:     **end while**
27:   **end for**

Algorithm 2 Maximum Channel Utilization Ratio [5]

Input: $U_{c_i}^m$

Output: ξ, χ
```
 1:  for i=1 to min{m,c} do
 2:      Select vehicles randomly until m
 3:      ξ ← max {U_{c_i}^m}
 4:      if m > c then
 5:          ξ ← 0 for vehicles which has not chosen
 6:          χ ← 0
 7:      end if
 8:      χ ← 1
 9:  end for
```

Alg. 2 is based on the single channel selection approach. Therefore, vehicles use the same channel throughout entire communication period. Channel utilization ratios, U_{ci}^m, are calculated for each vehicle as given in Eqn. 1. Here, each vehicle selects a channel with maximum utilization ratio in order to utilize the channel as long as possible. However, one of the drawback of single channel selection approach is that if the number of vehicle is higher than the number of available channels, all vehicles cannot access the channels. This situation causes a dramatic degradation of network performance, especially in high traffic density.

Algorithm 3 Checking Channel Utilization Map [5]

Input: $[U]_{cxt}$, $U_{c_i}^m$

Output: ξ, χ

Temp. Variables: t', γ
```
 1:  for i=1 to min{m,c} do
 2:      for j=1 to t do
 3:          t' ← Check the channel availability status
 4:          if t' is not empty then
 5:              Select a channel index, γ, randomly among available channels
 6:              ξ ← U_{c_γ}^m
 7:              χ ← 1
 8:          end if
 9:      end for
10:  end for
```

Alg. 3 is based on the single channel selection approach similar to Alg. 2. In this algorithm, vehicles check its utilization map by starting initial time slot. Whenever one or more channels are detected as available, a channel among available channels is selected randomly and this channel is used throughout all communication periods.

Algorithm 4 Checking Channel Utilization Table by Selecting Maximum Channel Utilization Ratio [5]

Input: $[U]_{cxt}$, $U_{c_i}^m$

Output: ξ, χ
Temp. Variables: t', γ
  1: **for** i=1 to min{m,c} **do**
  2:    **for** j=1 to t **do**
  3:      t' ← Check the channel availability status
  4:      **if** t' is not empty **then**
  5:        Select a channel index, γ, that has max utilization ratio among available channels
  6:        $\xi \leftarrow U_{c_\gamma}^m$
  7:        $\chi \leftarrow 1$
  8:      **end if**
  9:    **end for**
 10: **end for**

Alg. 4 uses single channel selection approach. In this algorithm, channel utilization ratios, U_{ci}^m, are calculated for each vehicle by utilizing utilization map and each vehicle checks its utilization map by starting from initial time slot similar to Alg. 3. Here, when one or more channels are detected as available at a time, the vehicle selects the channel that has maximum utilization ratio among available channels in this specific time and this channel is used throughout the transmission period.

Algorithm 5 Optimization Between Satisfaction Ratio and Channel Switching [5]

Input: $[U]_{cxt}$

Output: ξ, χ
Temp. Variables: γ, ε, μ
  1: Run algorithm 1 and obtain the satisfaction matrix [ξ], number of channel switching, χ in Algorithm 1
  2: **for** i=1 to m **do**
  3:    γ ← Specify beginning indices of transmission periods in [ξ]
  4:    ε ← Calculate transmission period in each channel switching according to γ
  5: **end for**
  6: **for** i=1 to m **do**
  7:    **for** j=1 to c **do**
  8:      μ ← Calculate the mean of transmission periods
  9:      **if** $\varepsilon < \mu$
 10:        $\varepsilon \leftarrow 0$
 11:        [ξ] ← Update satisfaction matrix according to ε
 12:        $\chi \leftarrow \chi - 1$
 13:      **end if**
 14:    **end for**
 15: **end for**

After single channel selection algorithms are defined in Algs. 2, 3 and 4, multi-channel selection algorithms will be defined in Algs. 5 and 6 in order to avoid frequent channel switching in Alg. 1. Therefore, in Algs. 5 and 6, at first Alg. 1 is run in order to obtain performance parameters, satisfaction matrix and the number of channel switching. Then the total amount of transmission intervals in satisfaction matrix is calculated in each channel switching as seen in example Figure 10. With the help of obtaining these transmission intervals in each channel switching, the mean of transmission periods is calculated so that transmission periods that smaller than the mean value are eliminated to decrease the number of channel switching in Alg. 1.

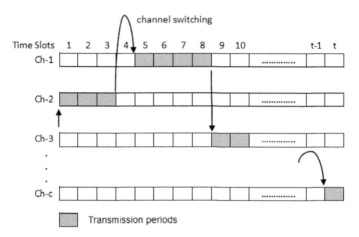

Figure 10. The structure of Algorithms 5 and 6 [5].

Algorithm 6 Assigning Number of Channel Switching by Vehicles [5]

Input: $[U]_{cxt}$, χ'

Output: ξ, χ
Temp. Variables: γ, ε

1:   **for** i=1 to m **do**
2:      Assign the number of channel switching, χ'
3:   **end for**
4:   Run algorithm 1 and obtain the satisfaction matrix [ξ], number of channel
     switching, χ in Algorithm 1
5:   **for** i=1 to m **do**
6:      $\gamma \leftarrow$ Specify beginning indices of transmission periods in [ξ]
7:      $\varepsilon \leftarrow$ Calculate transmission period in each channel switching according to γ
8:   **end for**
9:   **for** i=1 to m **do**
10:     **for** j=1 to c **do**
11:        sort ε values in descending order
12:        select maximum ε values in order until χ'
13:        [ξ] $\leftarrow$ Update satisfaction matrix according to ε
14:     **end for**
15:   **end for**

In Alg. 6, the number of channel switching is assigned by each vehicle. Similar to Alg. 5, Alg. 1 is run and transmission intervals are obtained. Then, depending on the assigned number, vehicles switch the channel by starting from maximum transmission period.

One of the important observations in Algs. 5 and 6 is that satisfaction ratios are decreased when compared to Alg. 1. However, while decreasing satisfaction ratios, the number of channel switching is also decreased and it is observed that satisfaction ratio and number of channel switching are optimized by conserving full network connectivity.

3.7 Results

In this section, cognitive channel selection algorithms are evaluated by showing the trade-off between network connectivity and channel switching. Connectivity problem is analyzed with two different approaches, with a queue and also without queue. The description of these approaches are defined as follows:

- *With queue:* In this approach, communication requests are modeled with M/M/m queuing model. If there is no available channel for communication requests at a time, requests are waited in the queue to serve next slots.

- *Without queue:* When a vehicle needs to access a channel, available channels are determined at this specific time period. However, if there is no available channel, communication requests are dropped.

In this respect, Figure 11 shows the relationship between satisfaction ratio of vehicles and the number of channel switching depending on traffic density. It is clearly showed that multi-channel selection algorithms (Algs. 1, 5 and 6) outperform when compared to single channel selection algorithms. Alg. 1 enables maximum connectivity with frequent channel switching. In Algs. 5 and 6, while the number of channel switching is decreased, it is observed that the satisfaction ratios are also decreased. However, simulation results show that satisfaction ratio and number of channel switching are optimized by conserving full network connectivity in Algs. 5 and 6 and maintaining continuous connectivity is more challenging in high traffic density.

Moreover, network connectivity is enhanced 26% with the queuing model. However, while enhancing connectivity, the number of channel switching also increases 12% with the queuing model. In particular, in a high traffic density, more vehicles spent more time to serve in the network and the opportunity to access the channels cannot be achieved at the end of the transmission range of RSU with the queuing model. Therefore, without queue approach can also be considered for connectivity modeling.

4. Summary

In this chapter, we analyze the network connectivity problem in CVNs. This chapter provides a taxonomy of the existing literature on network connectivity issues in CVNs. Maintaining a robust and continuous connectivity can be achieved with detecting spectrum opportunities and the presence or absence of primary users, deciding the

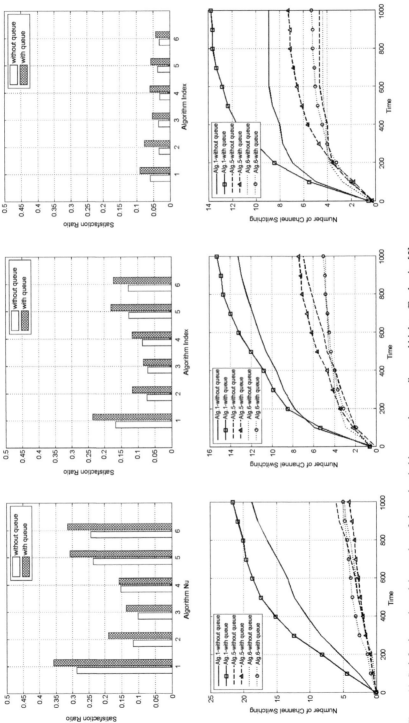

Figure 11. Satisfaction ratio and the number of channel switching w.r.t. low, medium and high traffic density [5].

best available channels by minimizing harmful interference with primary users, using the available spectrum resources in a fairly manner among vehicles. Therefore, the challenges of high vehicular mobility, the increasing use of vehicular applications, spatial and temporal changes of available channels and limited transmission range are described. Moreover, we give the details of an existing work for connectivity modeling in CVNs.

References

[1] I.F. Akyildiz, W. Lee, M.C. Vuran and S. Mohanty. Next generation/dynamic spectrum access/ cognitive radio wireless networks: A survey. Elsevier Computer Networks, 50(13): 2127–2159, 2006.

[2] I.F. Akyildiz, W. Lee, M.C. Vuran and S. Mohanty. A survey on spectrum management in cognitive radio networks. IEEE Communications Magazine, 46(4): 40–48, 2008.

[3] J.J. Blum, A. Eskandarian and L.J. Hoffman. Challenges of intervehicle ad hoc networks. IEEE Transactions on Intelligent Transportation Systems, 5(4): 347–351, 2004.

[4] J.A. Bondy and U.S.R. Murty. Graph theory with applications. Elsevier Science Publishing Co., Inc. New York, USA, 1976.

[5] E. Bozkaya and B. Canberk. Robust and continuous connectivity maintenance for vehicular dynamic spectrum access networks. Elsevier Ad Hoc Networks, 25(A): 72–83, 2015.

[6] E. Bozkaya, M. Erel and B. Canberk. Connectivity provisioning using cognitive channel selection in vehicular networks. Lecture Notes of the Institute for Computer Sciences, Social Informatics and Telecommunications Engineering, Ad Hoc Networks, 140: 169–179, 2014.

[7] S. Chen, R. Vuyyuru, O. Altintas and A. Wyglinski. On optimizing vehicular dynamic spectrum access networks: Automation and learning in mobile wireless environments. In IEEE Vehicular Networking Conference (VNC), Amsterdam, The Netherlands, 2011.

[8] S. Chen, A.M. Wyglinski, R. Vuyyuru and O. Altintas. Feasibility analysis of vehicular dynamic spectrum access via queueing theory model. Second IEEE Vehicular Networking Conference (VNC), Jersey City, New Jersey, USA, 2010.

[9] N. Cheng, N. Zhang, N. Lu, X. Shen,W. Mark and F. Liu. Opportunistic spectrum access for CR-VANETs: A game-theoretic approach. IEEE Trans. Veh. Technol., 63(1): 237–251, 2014.

[10] B.S.C. Choi, H. Im, K.C. Lee and M. Gerla. Cch: Cognitive channel hopping in vehicular ad hoc networks. IEEE Vehicular Technology Conference (VTC), San Francisco, United States, 2011.

[11] F. Cuomo, A. Abbagnale and A. Gregorini. Impact of primary users on the connectivity of a cognitive radio network. In Ninth IEEE Ad Hoc Networking Workshop (Med-Hoc-Net), Juan Les Pins, France, 2010.

[12] L. Dung and B. An. On the analysis of network connectivity in cognitive radio ad-hoc networks. In IEEE International Symposium on Computer, Consumer and Control, Taichung, Taiwan, 2014.

[13] M.D. Felice, K.D. Chowdhury and L. Bononi. Cooperative spectrum management in cognitive vehicular ad hoc networks. In IEEE Vehicular Networking Conference (VNC), Amsterdam, Netherlands, 2011.

[14] M.D. Felice, R. Doost-Mohammady, K.R. Chowdhury and B. Luciano. Smart radios for smart vehicles: cognitive vehicular networks. IEEE Vehicular Technology Magazine, 7(2): 26–33, 2012.

[15] M.M. Gad, A.A. Farid and H.T. Mouftah. A new connectivity metric for cognitive radio networks. In 24th IEEE International Symposium on Personal, Indoor and Mobile Radio Communications (PIMRC), London, United Kingdom, 2013.

[16] Y. Guo, Y. Ma, K. Niu and J. Lin. Connectivity of interference limited cognitive radio networks. In 3rd IEEE International Conference on Network Infrastructure and Digital Content (IC-NIDC), Beijing, China, 2012.

[17] Y. Han, E. Ekici, H. Kremo and O. Altintas. Throughput-efficient channel allocation in multichannel cognitive vehicular networks. Proceedings of the 33rd Annual IEEE International Conference on Computer Communications (INFOCOM'14), Toronto, Canada, 2014.

[18] J. Harri, F. Filali and C. Bonnet. Mobility models for vehicular Ad Hoc networks: A survey and taxonomy. IEEE Communications Surveys and Tutorials, 11(4): 19–41, 2009.

[19] X. Jin, S. Weijie and Y. Wei. Quantitative analysis of the VANET connectivity: Theory and application. 73rd IEEE Vehicular Technology Conference (VTC), Budapest, Hungary, 2011.

[20] M.J. Khabbaz, C.M. Assi and A. Ghrayeb. Modeling and analysis of DSA-based vehicle to-infrastructure communication systems. IEEE Transactions on Intelligent Transportation Systems, 14(3): 1186–1196, 2013.

[21] H. Li and D.K. Irick. Collaborative spectrum sensing in cognitive radio vehicular ad hoc networks: Belief propagation on highway. Proceedings of the 71st IEEE Vehicular Technology Conference (VTC), Taipei, Taiwan, 2010.

[22] J. Liu, Q. Zhang, Y. Zhang, Z. Wei and S. Ma. Connectivity of two nodes in cognitive radio ad hoc networks. IEEE Wireless Communications and Networking Conference (WCNC), Shanghai, China, 2013.

[23] D. Lu, X. Huang, P. Li and J. Fan. Connectivity of large-scale cognitive radio ad hoc networks. Proceedings IEEE INFOCOM, Orlando, FL, 2012.

[24] K. Menger. Zur allgemeinen Kurventheorie. Fundamenta Mathematicae, 10: 96–115, 1927.

[25] D. Niyato, E. Hossain and P. Wang. Optimal channel access management with QoS support for cognitive vehicular networks. IEEE Trans. Mobile Comput., 10(4): 573–591, 2011.

[26] S. Pagadarai, B. Lessard, A. Wyglinski, R. Vuyyuru and O. Altintas. Vehicular communication: Enhanced networking through dynamic spectrum access. IEEE Veh. Technol. Mag., 8(3): 93–103, 2013.

[27] S. Pagadarai, A. Wyglinski and R. Vuyyuru. Characterization of vacant UHF TV channels for vehicular dynamic spectrum access. IEEE Vehicular Networking Conference (VNC), Tokyo, Japan, 2009.

[28] S. Rocke, S. Chen, R. Vuyyuru, O. Altintas and A. Wyglinski. Knowledge based dynamic channel selection in vehicular networks. IEEE Vehicular Networking Conference (VNC), Seoul, Republic of Korea, 2012.

[29] C. Silva, E. Cerqueira and M. Nogueira. Connectivity management to support reliable communication on cognitive vehicular networks. IEEE Wireless Days (WD), 2014 IFIP, Rio de Janeiro, Brazil, 2014.

[30] M. Sultana and K.S. Kwak. Non-preemptive queueing-based performance analysis of dynamic spectrum access for vehicular communication system over tv white space. Third IEEE International Conference on Ubiquitous and Future Networks (ICUFN), Dalian, China, 2011.

[31] K. Tsukamoto, Y. Omori, O. Altintas, M. Tsuru and Y. Oie. On spatially aware channel selection in dynamic spectrum access multi-hop inter-vehicle communications. 70th IEEE Vehicular Technology Conference (VTC), Anchorage, Alaska, USA, 2009.

[32] X.Y. Wang and P. Ho. A novel sensing coordination framework for CR-VANETs. IEEE Transactions on Vehicular Technology, 59(4): 1936–1948, 2010.

[33] H. Whitney. Congruent graphs and the connectivity of graphs. Amer. J. Math., 54: 150–168, 1932.

[34] Non-separable and planar graphs. Trans. Amer. Math. Soc., 34: 339–362, 1932.

[35] D. Helbing. Traffic and related self-driven many-particles systems, Rev. Modern Physics, 73: 1067–1141, 2001.

[36] S. Rathore, M. Naiyar and A. Ali. Comparative study of entity and group mobility models in MANETs based on underlying reactive, proactive and hybrid routing schemes. IEEE International Symposium on Communications and Information Technologies, Bangkok, Thailand, 2006.

[37] E.R. Cavalcanti and M.A. Spohn. Predicting mobility metrics through regression analysis for random, group, and grid-based mobility models in MANETs. IEEE Symposium on Computers and Communications (ISCC), Riccione, Italy, 2010.

Part III

Applications for CVNs

8

Security and Privacy in Vehicular Networks
Challenges and Algorithms

Yi Gai,[1], Jian Lin[2] and Bhaskar Krishnamachari[3]*

ABSTRACT

By connecting the vehicles and roadside units (RSUs), vehicular networks (VANETs) enable promising applications for enhancing road safety, mitigating traffic and disseminating safety information among drivers and passengers. The advancement of cognitive radio (CR) technologies, including dynamic spectrum access and adaptive software-defined radios, can improve vehicular communication efficiency by achieving more efficient radio spectrum usage. The prevalent deployment of such vehicular networks, as well as the employment of cognitive radio technologies, raises important concerns about its security and privacy. We must design security mechanisms that ensure integrity, confidentiality, system performances and robustness against frequent and severe malicious attacks.

In this survey paper, we characterize security and privacy issues in vehicular networks with cognitive radio technologies and discuss solutions in achieving secure cognitive communications in vehicular networks. We review how security threats have been raised along with new architectures of vehicular networks. Based on the understanding of potential security threats, we present some current

[1] Intel Labs, Hillsboro, OR 97124, USA.
[2] Department of Electrical and Computer Engineering, Georgia Institute of Technology, Atlanta, GA, 30332, USA.
    Email: jlin61@gatech.edu
[3] Department of Electrical Engineering, University of Southern California, Los Angeles, CA 90089, USA.
    Email: bkrishna@usc.edu
* Corresponding author: yigaiee@gmail.com; yi.gai@intel.com

methods that provide security services and preserve privacy. Although these methods can be leveraged to address some security problems, they do not provide a full solution. In fact, many unique challenges exist in new vehicular network architectures with cognitive radio, and remain to be addressed. We conclude the paper with a few promising future research directions towards ensuring security and privacy in vehicular networks.

1. Introduction

Vehicular communications have emerged as a new technology towards creating safer and more efficient driving conditions. By allowing short-range direct communications between vehicles (vehicle-to-vehicle, V2V) and roadside units (vehicle-to-infrastructure, V2I), vehicular ad hoc networks (VANETs) enable promising applications for enhancing road safety, mitigating traffic and disseminating safety information among drivers and passengers [1, 2]. Numerous efforts on VANETs development have been made by governments, telecommunication industry and automobile manufacturers paving the way for realizing the intelligent transportation systems (ITS) [3], including the IEEE 802.11p standard operating in the licensed ITS band of 5.9 GHz (5.85–5.925 GHz) for the north America. As a special type of mobile ad hoc networks (MANETs), VANETs tend to have high node density, high mobility, unbounded network size, and time critical missions. In light of the distinct characteristics of VANETs, low spectrum utilization reported for the licensed users (a.k.a. primary users) [4] will significantly compromise the effectiveness and efficiency of vehicular communications. To efficiently utilize the fallow spectrum, cognitive radio (CR) [5, 6] provides the key to enable dynamic spectrum access through spectrum sensing. The unlicensed (secondary) users in CR networks work on the opportunistic basis accessing the unused frequency band called "white space" or "spectrum hole" for communications without disturbing the primary users. Such a cognitive radio has both the cognitive feature and the reconfiguration capability, and is well suited for an implementation in the software defined radio (SDR) [5]. With the ever-growing interest in the CR system, it is envisioned that future vehicles will be CR-enabled [7] and CR-VANETs, as illustrated in Figure 1, will have a highly potential market due to the huge consumer size and the myriad applications.

Along with all the benefits that vehicular networks offer, security and privacy are the key hindrances to the wide spread implementation, because an insecure and unreliable VANET can be more disturbed than a system without it. Frequent and malicious attacks to the vehicles and roadside infrastructure can incur financial loss and even loss of human life [2], for example, from accidents caused by tampered interaction between vehicle radios and the traffic signal infrastructure. The road traffic can even be disabled by a relatively unsophisticated attacker. Consider, for example, a single compromised vehicle disseminating falsified hazard warnings can disrupt all vehicles in bidirectional traffic streams. Therefore, reliable security measures should be taken to assure that the messages are generated by a trusted source, and the data packets are not tampered with before they are received at an authenticated receiver. In

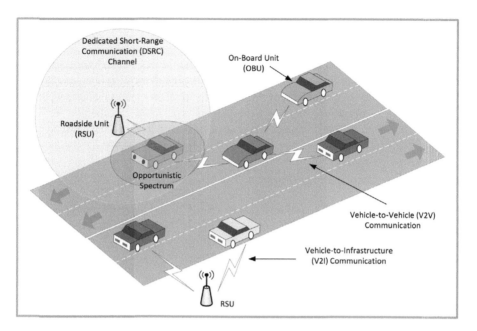

Figure 1. Basic CR-VANET network model.

addition, the unique characteristics of the CR-enabled VANETS such as the cognitive features and the high-speed mobility pose difficult challenges and spawn new types of security threats.

Important concerns on the privacy in vehicular networks have also been raised. Due to the broadcast nature of the wireless channel, safety and non-safety related messages can be eavesdropped by another entity within the radio ranges. Such messages can include private information about the vehicle such as location and speed, which can be collected, analyzed and misused by a malicious entity. When the location of a vehicle is tracked, it is easy to build the vehicle's profile and thus breach the driver's privacy. In addition, since the vehicle's identifiers such as an electronic license plate (ELP) is often mapped to the identity of its driver, the personal data of the driver can be possessed by an unwanted party through identity revealing. Therefore, it is a prerequisite to provide driver privacy so that the true identity of the vehicle sending each data packet is kept private or immune of misuse. Without a trustful privacy measure, drivers who are concerned about tracking and identity revealing might simply disable the vehicle radio and consequently disable all beneficial applications of the VANETS.

The rest of this chapter is organized as follows. In section 2, we discuss the challenges, security requirements and the possible attacks for vehicular networks. Section 3 presents the threats/possible attacks for cognitive radio technologies, including the threats related to spectrum sensing, software defined radio, and the link layer. Section 4 concludes the paper and points out the future directions.

2. Security, Privacy and Safety in Vehicular Networks

Similar to classical networks, many attacks are threatening the VANETs. Security and Privacy issues are crucial in VANETs as the potential threats can be harmful to traffic flows and even human lives. For example, severe traffic disruption and accidents can be caused by fake messages. There are many mechanisms being developed to fend off many types of attacks and ensure the efficient operation of vehicular networks. Some of the security and privacy challenges in VANETs have been discussed in [8, 9, 10, 11, 12, 13].

2.1 Vehicular Network Challenges

Security and privacy are highly important in VANETs, and there are many challenges remaining to be addressed. Security protocols should meet the specific characteristics of VANETs such as high mobility, volatility and network scalability [14, 8, 10].

- *High mobility.* One of the most important features of VANETs is the high mobility of nodes, which introduces many challenges on security. Communications between nodes in VANETs are direct, without relying on any particular infrastructure. Each node in the network is mobile, and can move from one place to another very quickly within the communication coverage area, or even out of the coverage range. The connections between two nodes may last a very short period of time as they will never meet again, and some other intermediate nodes should act as routers since no dedicated routers are deployed. The high mobility of the nodes not only puts considerable challenges in the protocol design, it also places many problems in securing the network. For example, the high mobility of the nodes has made classical authentication not suitable for VANETs. Given the fact that many nodes may only communicate once, classic mechanisms such as handshake are impossible in a practical protocol.

- *Volatility.* Due to the high mobility, the connectivity between two nodes can vary quickly and switch between on and off frequently. Vehicles may also travel in the opposite direction so the connections between them will be ephemeral. Using long time passwords to secure users' devices is impractical due to the lack of long live context and stable connections in VANETs [15].

- *Network scalability.* As VANETs are getting more and more popular, the number of nodes is growing very quickly [16]. So the scalability of the network has put many challenges in the implementation and deployment of such a large ad hoc network. The management of identifications and distribution of public and private keys become very challenging as the network size grows, indicating the scalability problem. Also, security and privacy issues change in different areas in the world, so it is impossible to use the same rules everywhere.

2.2 Security Requirements

It is important to address the security requirements that the system must respect before addressing the security issues. Possible security threats may be caused by failing to respect a requirement. Below are a lot of security requirements that should be met [14, 8].

- *Authentication.* Authentication plays a crucial role in securing the vehicular network, and is one of the key requirements. Important information, such as the identification, location and property of the message sender, needs to be secured in vehicular networks. It is important to have ID authentication to ensure that the communicating nodes are legitimate. This authentication allows a node to identify the transmitter of received messages in a unique way. It is also important to have data authentication to ensure that the transmitted message is what it is declared to be. Authentication in vehicular networks can prevent some attacks such as Sybil attacks mentioned in Section 2.3.2.

- *Availability.* For vehicular networks, some data are required to be transferred in real time for making time critical decisions at the end users. VANETs should remain operational even when faults, or malicious/critical conditions occur, or during the evolvement of the roadside infrastructure. This requires a secure, fault-tolerant and scalable network design that can handle the desired and predicted network load with any possible attacks, such as Denial-of-Service (DoS) attacks (see Section 2.3.1). Cognitive vehicular networks are more likely to be harmed by DoS attacks since the wireless transitions make legitimate or malicious users easier to access and exploit the resource.

- *Non-repudiation.* Non-repudiation ensures that the acknowledgement from vehicles on sending messages to any destined vehicle. Digital signature can be used to sign all messages before the transmission. It is necessary to employ non-repudiation for preventing legitimate users from denying the transmission of their messages. ECC [17] is one of the main encryption method used by VANETs due to its good performance and complex cryptographic scheme. A good scheme to send and verify signatures of messages is the combination of ECC and non-repudiation of origin. Non-repudiation of origin can help with the identification of attackers who attempt to send harmful information since it enables every vehicle to be accountable for the messages generated.

- *Privacy.* As the exchange of data in VANETs has created great analytical opportunities, privacy violations have become a crucial problem [14]. Most drivers want to keep their personal information private and no other unauthorized users should know about it. Such information includes the name of the driver/ user, speed of the vehicle, internal sensor data, ELP, etc. Keeping users' present and past location private is also crucial in VANETs, since an attacker or any unauthorized party should not track a vehicle's location. The location of vehicle should also be kept private so that no attacker can find the location or the route of any particular users. The design of VANETs' architecture should respect users' privacy and protect them from malicious nodes at the same time. This means users' anonymity when communicating with each other should be allowed while respecting the trusted base system of VANETs. Another challenge here is that some sensitive information such as vehicular location, identity, speed is sent via broadcasts over wireless communications, so such information can be received by any user in the network and should be out of reach of attackers or unauthorized receivers.

- *Access control.* Rights and privileges that each nodes in the network can perform are defined by access control. It also defines the operations of each node to specific services provided by the infrastructure or other nodes through local policies. Access control should ensure any misbehaving nodes to be revoked and sensitive communications such as messages from police cars not be heard by other nodes without authorization. It prevents unauthorized nodes from accessing the service that they are not allowed to access. The rights of each node are established through authorizations, which are part of access control [18].

- *Integrity.* Data integrity prevents authorized creation, modification, and alteration of the data. It ensures and maintains the accuracy and consistency of data in VANETs. It requires that data should not be modified or altered during transmissions and the received message must match the message sent over the air.

2.3 Possible Attacks

2.3.1 Denial of Service (DoS) Attacks

Denial of Service attack is a type of attack on a network that bring the network down by overwhelming it with traffic from multiple sources. DoS attack can be classified into two categories when it is achieved by overwhelming the resources [19]: DoS attack in V2V communications and DoS attack in V2I communications. DoS attack can be performed by jamming the networks, and this includes [20]: (1) trivial jamming attack, for which an attacker constantly inserts noise; (2) periodic jamming attack, for which a short signal is transmitted by an attacker periodically, and often enough to affect other transmissions; (3) reactive jamming attack, for which an attacher initiates the jamming whenever it detects a transmission between other nodes to cause a collision for the transmission before it ends.

DoS attack can harm the network by preventing the delivery of critical information. In the case when driver depends heavily on the information from application, it increases the danger and can raise serious problems. For example, a heavy traffic jam can be created by an attacker by preventing the warning of approaching vehicles. Figure 2 illustrates an example of DoS attack.

DoS attack has drawn the attention of many researchers. [21] proposed a simple real-time method to detect DoS attacks in VANETs, with a focus on the jamming attack on the periodic position message exchange. The method is validated by showing the attack detection and false alarm probabilities. In a recent work [22] by Verma and Hasbullah, a Bloom-filter-based detection method was proposed to detect and defend against the IP spoofing of addresses in the DoS attacks in VANETs. The scheme provided a simple and efficient end-to-end solution with the availability of a service for legitimate nodes in the network. The scheme can also work for high attack rates.

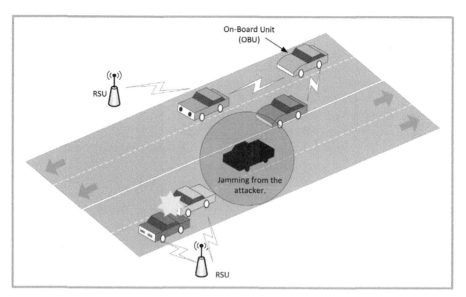

Figure 2. An example of DoS attack.

2.3.2 Sybil Attacks

An attacker claims multiple different identities, called Sybil nodes [23], to send multiple messages, for which different identities are used at the same time. As illustrated in Figure 3, this creates an "illusion" of traffic congestions such that other drivers are forced to take an alternative route and a clear path is left for the attacker to its

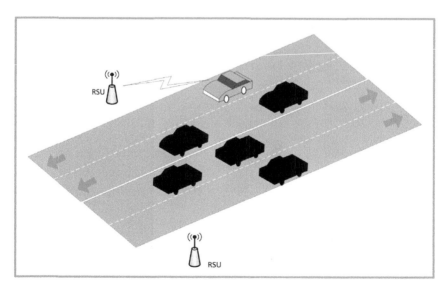

Figure 3. An example of Sybil attack.

destination. This attack depends on whether the system accepts inputs from entities without linking from a trust entity. It also depends on how easily identities can be generated, and whether all entities are treated the same by the system.

In [24], a secure and efficient protocol was proposed to ensure both message authentication and privacy preservation, with the ability to prevent Sybil attacks. The scheme is based on elliptic curve digital signature algorithm (ECDSA) with which secure signatures are generated to be used by nodes. Temporary identities are also generated using cryptographic techniques. Besides preventing Sybil attacks, the scheme can also detect several possible attacks such as DoS attacks and Grey Hole attacks.

2.3.3 Message Suppression/Alteration/Falsification Attacks

In message suppression attack [25], one or more vehicles may be used by an attacker to selectively drop packets. For example, the congestion avoidance warning may be removed by a prankster to prevent vehicles from selecting an alternative route and force the vehicles to wait in the traffic. The alteration/falsification attack [12] aims to alter the legitimate messages. An attack may delay the transmission of a message intentionally, or alter an individual message to indicate false information. For example, a message of a normal traffic may be altered by an attacker to falsely indicate a heavy traffic ahead.

The message suppression/alteration/falsification attacks can easily cause the car accident, traffic jam and the decrease of VANET performance in terms of bandwidth/resource utilization. A mechanism to defend against this type of attacks is to rely on the same information got from nodes which have not been hacked. [26] proposed a mechanism to check the message plausibility by adopting message overhearing mechanism with reputations. This helps verify the message in the network to ensure if a message is legitimate.

2.3.4 Other Attacks

Besides the above mentioned attacks, there are many other types of attacks that are threatening VANETs:

- *Social attacks*. The main objective of social attack [11] is to indirectly create problems in the network. For example, the attacker can intentionally cause drivers in the network with anger by broadcasting unappropriate message such as "You are an idiot!". Then the affect driver might cause other problems while driving.
- *Brute-force attacks*. As the name suggests, this attack includes the attack that an attacker uses brute-force technique to break cryptographic keys used in secure communications.
- *Replay attacks*. In this attack, attacker captures the previously generated messages and uses it in new connections of other parts of the network.
- *GPS spoofing*. The attacker alters the GPS information so that those drivers who rely on GPS information would be confused by a wrong location.

- *Timing attacks.* The purpose of this attack is to alter the time slot (e.g., add some time slot) for the transmitted message, so that some messages with critical information would be delayed and received after the required time.

[27] presented two approaches to prevent replay attacks: (1) using a globally synchronized time for all nodes and (2) using nonce. The first option will require a lot of efforts on the organization, so the second one is preferred. [28] proposed another solution to use the GPS information come with the vehicles. [29] presented a new scheme for malicious behavior prediction. Based on that, a reliable intrusion detection and prevention scheme was proposed for VANETs within a game theoretical model. A set of detection and identification rules were proposed for identifying malicious behavior, and thus preventing from the most dangerous attacks.

3. Security and Privacy for Cognitive Radio Technologies

In this section, we review and discuss security threats and the countermeasures for cognitive radio (CR) technologies. Considering that future vehicles are envisioned to be CR-enabled and that software defined radio (SDR) is a promising platform for implementing CRs, we elaborate our discussions in the following scope: the spectrum sensing, the SDR platform, and the link layer.

3.1 Threats Related to Spectrum Sensing

Threats related to spectrum sensing target mainly two fundamental characteristics of cognitive radios, namely cognitive capability and reconfigurability. The major threats are primary user emulation (PUE) attack and objective function attack.

3.1.1 Primary User Emulation Attacks

When a primary user is detected in a spectrum band, the secondary user must vacate the band by performing a possible hand-off to another spectrum hole. In primary user emulation (PUE) attack [30, 31], a malicious secondary user masquerades as a primary user to gain exclusive access to the given band. For less sophisticated attackers, this can be achieved by emitting RF energy at the primary user's frequency. For more sophisticated attackers, this can be achieved by emulating the characteristics of the primary signal such as the cyclostationary characteristics. Two motivations behind the PUE attacks are: selfish motivation and malicious motivation. The selfish motivation is to gain full usage of the band preempting other secondary users' access. The malicious motivation is to cause the operation disruption, i.e., denial-of-service (DoS). In particular, malicious users can pollute multiple spectrum holes to cause extensive DoS.

Depending on whether the location of the primary user (PU) is known a priori, detection of PUE can be categorized into location-based schemes and non-location based schemes. Location-based schemes rely on the known location of PU, such as the fixed TV tower, and an estimation of the location for the PU or an attacker [32]. The PUE attack is detected by comparing the estimated and the true locations. The location information is estimated using the received signal strength (RSS) in [32], and the time difference of arrival (TDOA) and the frequency difference of arrival (FDOA) in [33]. A hypothesis testing algorithm for PUE detection is given in [34] using Wald's sequential probability radio test (WSPRT), assuming the attacker uses fixed transmission power. Considering multiple attackers and the fading effect of the wireless channel, an algorithm based on Neyman-pearson composite hypothesis test (NPCHT) is presented in [35]. The scenario where attackers use variable power levels is handled in [36]. These location-based methods suffer from the volatility in parameter estimation. Also, location information is insufficient in the mobile network context where the PUs are mobile and power conservative. On the other hand, non-location based schemes utilize signal feature detection, for example, cyclostationary feature detection and matched filter detection, to capture the distinctive characteristic of the primary signal. Initially, RF fingerprinting (RFF) has been proposed as a mean of enhancing wireless security in [37], which identifies the emitter by monitoring and analyzing the analog signal at the physical layer. A cross-layer signal pattern recognition technique was introduced in [38] that explores the unique property call electromagnetic signature (EMS, which can be compared with the human biometric feature) of each CR device to build a security sub-system. The malicious device can be detected based on its signal pattern within certain levels of deviations. Relying solely on signal feature detection may not be sufficient as a sophisticated attacker can mimic several characteristics of the primary signal. Also, the storage requirement and sensing time tend to increase due to the overhead of signal processing operation.

3.1.2 Objective Function Attacks

The reconfiguration capabilities of cognitive radio allow it to learn from the past and make intelligent decisions to adjust the radio parameters to achieve optimal performance. The parameters, such as power, center frequency, modulation and coding, and medium access protocol, are adjusted by the cognitive engine to satisfy a multi-objective function. An attacker can launch the objective function attack by manipulating the spectrum environment during the learning phase of the CRs and thus affecting the learning beliefs of the CRs. Consequently, the CRs' behavior is fitted into a manipulated level of optimality. An example is given in [39], where a radio has three goals: low-power, high-rate, and security. The objective function is expressed as $f = w_1P + w_2R + w_3S$, where P, R and S represent the level of power, rate and security, respectively. Assuming the attacker wanted to force the CR to use a lower security level $s_1 < s_2$, it reduces the data rates (from r_2 to r_1) of the CR by jamming the channel whenever the CR uses a higher security level. As long as this artificial interference causes the objective function to decrease, i.e., $w_1P + w_2r_1 + w_3s_2 < w_1P + w_2r_2 + w_3s_1$, the CR would never use the higher security level s_2. Similarly, an adversary can manipulate the CR's learning belief on a variety of parameters, as long as they are components of the objective function. It is noted that objective function attack works only when the CR is performing online learning.

To defend against objective function attack, a simple threshold based scheme is suggested in [40] to invalidate the updated radio parameter that does not meet the threshold. In [41], the authors proposed a multi-objective programming (MOP) model to allow CR to adapt to surrounding intelligently. In this model, the CR compares the parameters with fitness value with MOP to decide whether attackers exist. If so, the CR adjusts the tampered parameters to the optimal settings.

3.2 Physical Radio Attacks

3.2.1 Software Defined Radio

The software defined radio (SDR) and cognitive radio have emerged as revolutionary techniques. SDR technology implements radio functionalities that provide high degree of flexibility and reconfigurability. An example is shown in Figure 4, in which the left half represents the universal software radio peripheral 1 (USRP1), and the right half represents the software components such as GNU radio. The USRP1 is a hardware platform that allows a software radio to be implemented in a general purpose processor, which consists of a radio frequency (RF) daughter board that functions as an RF front-end and a main board that has an analog-to-digital converter (ADC), a digital-to-analog converter (DAC), a field-programmable gate array (FPGA), and universal serial bus (USB) interfaces.

The GNU radio (and similar software architecture) provides various applications and waveform processing blocks.

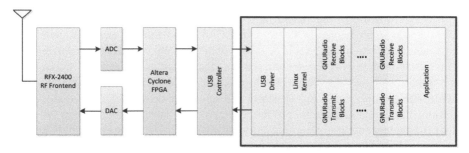

Figure 4. Software defined radio (SDR) blocks: the left half represents USRP hardware components, and the right half represent software components.

3.2.2 SDR Threats

The SDR security is of paramount importance, because with a compromised SDR an attacker can launch combinations of attacks affect all parts of the cognitive cycle. The main security threats to SDR are [42]: (1) software attack, and (2) unauthorized use of SDR services. These threats are introduced during either the installation of the software or the operation of the software [43]. The software attack refers to attacks in any of the software components including real-time operating system, the software framework, waveforms, and applications. The unauthorized usage of SDR services takes place when a waveform or application can access or use services of the SDR platform for which it is not certified to. For example, with tampered software that controls the physical layer, the attacker can force the RF module to output signals that do not satisfy the local radio regulation parameters. The "device cloning" threat referring to unauthorized use of services of another SDR service is described in [44]. It has also been shown that it is very difficult to detect security holes or Trojan horses (e.g., unauthorized software) in the product testing phase [43].

The countermeasure against software attacks have been proposed from the perspectives of hardware, software and middlewear. In [45], a secure software downloading system is proposed that uses the characteristic of the field programmable gate arrays (FPGAs) composing the SDR. The structure of a FPGA consists of independent configuration logic blocks, which allows systems to be arranged in a variety of different layouts and thus enables high security encipherment. By numerical evaluation, the authors show that the proposed system has high immunity to illegal acquisition of software using replay attack. In [46], the authors proposed a light secure socket layer (LSSL) to securely connect manufacture's server and SDR devices. Comparing with SSL, LSSL consumes less bandwidth and is more suitable for low-bandwidth low-capability SDR devices. The secure protocol in [46] also includes mutual authentication, public/private key mechanisms for data encryption and decryption, and fingerprint calculation for data integrity. However, the tamper-proof hardware approach in [45] cannot adapt the security policy to the changing environment, and software security module in [46] may itself be vulnerable to malicious modification/blocking. To mitigate these limitations, the authors in [47] proposed a secure radio middleware (SRM) that is based on robust separation of the radio operation environment and user application environment through the use of visualization. A prototype is implemented in VMware and GNU radio toolkit, assuming generously available resources. Cares should be taken when this architecture is applied to resource limited devices. In [44], the authors presented an anti-cloning framework that uses a set of hardware and software technologies. The SDR mobile device works together with wireless operator (WO) to detect if it is a cloned or valid unit, where WO is assumed to be the manufacture of the SDR. Another advantage of the framework is that it is independent of communication protocols and works well for both cellular technologies and the internet.

3.3 Link Layer Attacks

In practice, the wireless channel effects such as shadowing and multipath fading, as well as receiver uncertainty can significantly compromise the detection performance of spectrum sensing for an individual CR. To mitigate these issues, cooperative sensing [48] explores spatial diversity from observations of spatially separated CR users. Based on how cooperating CR users share their sensing data in the network, cooperative spectrum sensing can be classified into three categories: centralized, distributed and relay-assisted. For all categories, secondary users (SUs) transmit their observations to the neighbors or a fusion center, which then fuses the collected data and makes the final decision on the presence or absence of an active primary user. How the SUs access the shared channel to report sensing data is managed at the link layer.

3.3.1 Spectrum Sensing Data Falsification (SSDF)

In spectrum sensing data falsification (SSDF), also know as Byzantine attack, the detection performance is compromised by falsified data sent from an attacker [49]. The SSDF targets centralized, as well distributed and relay-assisted CRNs. The malicious user can send a detection report when the primary user (PU) is absent and vice versa. With the false local spectrum sensing data, a "false-alarm" decision denies legitimate SUs from accessing a spectrum hole, and a "miss" decision causes significant interference to the active PU. There are three types of adversaries sending falsified sensing data: (1) malicious nodes that aim at causing interference to the PU or denying access of SUs, (2) greedy nodes that aim at maximizing ones' own goods, and (3) unintentional nodes that occasionally send inaccurate data due to the wireless channel effects (fading, shadowing, etc.), mobility, and the hidden node problem.

Advanced data fusion techniques are required to defend against SSDF attack. Specifically, the data fusion mechanism must discriminate all sensor terminals with their trust or reputation level. Ideally, only the data from reliable sensing devices should be accepted while unreliable devices should be filtered. In [50], a two-level defense against SSDF attack, called weighted sequential probability ratio test (WSPRT), is proposed. The first step is a reputation maintenance step, and the second step is the actual hypothesis test, in which the data collector authenticates sensing reports. The purpose of the first level is to prevent relay attacks or falsified data injected from outside of the CR network. All CRs have an initial reputation value of zero, which increments upon each correct local data report. The hypothesis test step is based on sequential probability ratio test (SPRT), which is a hypothesis test for sequential analysis and supports sampling a variable number of observations. The SPRT takes into account each CR's reputation value. The WSPRT method has several drawbacks, which include the requirement of a large number of samples and thus long sensing time, the possibility of leading to a deadlock with no decision made, and low performance in a highly dynamic environment. To overcome these drawbacks, the authors in [51] presented two enhanced schemes, named enhanced weighted sequential probability ratio test (EWSPRT) and enhanced weighted sequential zero/one test (EWSZOT). Specifically, EWSPRT adopts a more aggressive weight module while using the same test module as WSPRT. EWSZOT uses the same aggressive weight module as EWSPRT and a new sequential 0/1 test module instead of SPRT. It is shown through simulation that the sampling numbers of EWSPRT and EWSZOT are 40% and 75% lower than WSPRT [50], with comparable errors rates.

In [52], the authors proposed a malicious user detection algorithm that calculates the suspicious level of SUs based on their past reports. The algorithm calculates the suspicious values as well as the consistency values to eliminate the malicious users' influence on the primary user detection results. The scheme is able to quickly reduce the trust value of a good user when this user suddenly turns bad. While the simulation results show significant performance advantage, the algorithm can only tackle one malicious user, and assumes the prior statistic knowledge for existence of the primary user. For the case of multiple attackers, in [53], the authors proposed a detection method, which counts mismatches between their local decisions and the global decision at the FC over a time window, and then removes the Byzantines from the data fusion process. The major drawback of this method is that unintentionally malicious users (due to temporary reasons) can not be restored.

3.3.2 Common Control Channel Threats

The common control channel (CCC) is an in-band or out-of-band channel, which is used by the SUs to exchange control information. The control information includes but is not limited to cooperative sensing, channel negotiation, and spectrum hand-off. When many SUs attempt to access the CCC at the same time, the channel becomes congested. A malicious user can jam the CCC in the first place to prevent the information exchange of SUs. It is shown in [54] that the CCC attack can lead to a near zero throughput for the SUs. Since the CCC can create a single point of failure, it is crucial to design CCC scheme resilient to control channel jamming attack.

It is noted that the single-hop networks can be made immune to CCC threats with an additional security sublayer, whereas multi-hops are more vulnerable to the CCC threats. In a single-hop network, such as IEEE 802.22 [55], a security sublayer is implemented to provide authorization, authentication and encryption. However, implementing such a security sublayer is very challenging in multi-hop networks. The dynamic CCC allocation methods have been proposed to combat CCC jamming attack by dynamically allocating CCC to maintain control communication. The dynamic CCC allocation is achieved by cross-channel communication in [56], which allows the CRs to transmit on the jammed channel as long as the receiver's channel is jamming free. The benefit of this method is the elimination of the need for the transmission pair to be both interference free, however, the cost of channel switching is high. A randomized distributed channel establishment scheme based on frequency hopping is proposed in [57] for mitigating CCC jamming in cluster-based ad hoc networks. Under this scheme, nodes in the network are able to temporarily construct a control channel until the jammer is removed from the network. The cluster head (CH) determines the hopping sequence and thus reduces the affected area to the cluster. To defend against the case that the CH is compromised, rotation of CHs should be considered. Another class of anti-jamming methods is based on CCC key distribution, which is polynomial-based in [58] and randomly distributed in [59]. In these approaches, the CCC keys are assigned to mask the CCC allocation in time slots with duplicate control transmissions on multiple CCCs. In [60], the authors provided a CCC security framework, in which an authentication phase is followed by encrypted transactions for channel negotiation between the transmitter and the receiver. While eavesdropping and unauthenticated access can be prevented under such a framework, the threats from compromised CRs that may manipulate the control data still exist. Moreover, it is noted that implementing the schemes in [58, 59, 60] is challenging in multi-hop networks, as there does not exist a trusted entity to distribute key materials.

4 Conclusion

This paper has surveyed security and privacy challenges in VANETs with the use of cognitive radio technology. Although CR-VANETs are still in its early stage, it is growing quickly with a lot of interests due to a great potential market for vehicular communications. The security challenges and requirements of CR-VANETs were discussed. Then, possible attacks and privacy and security solutions in both the vehicular architecture, cognitive sensing, physical radio, and link layer were presented. With the deployment of cognitive radio technologies for vehicular networks, security and privacy solutions need to adapt to its unique characteristics.

The main conclusions and recommendations for future directions are listed below.

- *Large scale capture attack.* Existing countermeasures can only detect and defend against a few malicious users. When a large number of users are compromised, they all report falsified data, existing methods such as sequential probability ratio test based cannot guarantee the detection performance. More advanced algorithms are needed to tackle large scale capture attack.

- *Key distribution design and infrastructure security.* While cryptography based framework provides robust authentication and integrity, implementing such a system in vehicular networks is challenging. An infrastructure such as the roadside unit could be the certification authority but at the same time can be a single point of failure. More studies are required to implement reliable and low-overhead cryptographic primitives.

- *Mobility parameters evaluation.* Among others, the high mobility distinguishes VENETs from other mobile ad hoc networks, therefore it is important to analyze and experimentally study the impact of mobility parameters on the performance of aforementioned countermeasures. The mobility parameters can include, for example, speed, direction, and profile of the cognitive radio enabled vehicles.

- *Cross-layer trust mechanism design.* Though trust mechanisms have been separately proposed for the MAC and routing layer, there lacks a joint design that guarantees trustworthiness down from the physical sensing up to the network level. In addition, more intelligent trust mechanism can be introduced in the learning process of the cognitive radio to defend against objective function attack.

- *Conflict of security and privacy.* Ensuing security and privacy at the same time for vehicular networks is a difficult challenge. This is in part because the needs of security and privacy are often in conflict. For example, the accountability is ensured by digital signature, which in turn compromises the anonymity. A careful system design is required to achieve the reasonable balance between security and privacy tailored to different applications.

- *Lower overhead of security policies.* Many current security solutions have brought in great overhead to establish and maintain the secure communications. The network overhead must be optimized especially when the network is experiencing congestion when data traffics are injected in the network.

- *Testing platforms.* Although many different solutions are developed to secure the vehicular networks, we lack of an effective testing platform that can be used to perform testing with different components of CR-VANETs. A platform used to directly test potential solutions to CR-VANETs issues can be very helpful to verify potential scalable solutions under different settings especially for a large network.

- *Formal methods for reliable software.* Formal techniques are very helpful with producing reliable software/applications for vehicular networks. For VANETs, we need a formalism to take into account of not only the many variables (such as positions, destinations, certificates), but also the users and users' behaviors.

References

[1] H. Hartenstein and K.P. Laberteaux. A tutorial survey on vehicular ad hoc networks. IEEE Communications Magazine, 46(6): 164–171, June 2008.

[2] Y. Toor, P. Muhlethaler and A. Laouiti. Vehicle ad hoc networks: applications and related technical issues. IEEE Communications Surveys Tutorials, 10(3): 74–88, Third 2008.

[3] L. Figueiredo, I. Jesus, J.A.T. Machado, J.R. Ferreira and J.L.M. de Carvalho. Towards the development of intelligent transportation systems. IEEE Intelligent Transportation Systems, pp. 1206–1211, 2001.

[4] G. Staple and K. Werbach. The end of spectrum scarcity spectrum allocation and utilization. IEEE Spectrum, 41(3): 48–52, March 2004.

[5] J. Mitola and G.Q. Jr. Maguire. Cognitive radio: making software radios more personal. IEEE Personal Communications, 6(4): 13–18, Aug 1999.

[6] S. Haykin. Cognitive radio: brain-empowered wireless communications. IEEE Journal on Selected Areas in Communications, 23(2): 201–220, Feb 2005.

[7] M. Di Felice, R. Doost-Mohammady, K.R. Chowdhury and L. Bononi. Smart radios for smart vehicles: Cognitive vehicular networks. IEEE Vehicular Technology Magazine, 7(2): 26–33, June 2012.

[8] J. Lacroix and K. El-Khatib. Vehicular ad hoc network security and privacy: A second look. In The Third International Conference on Advances in Vehicular Systems, Technologies and Applications (VEHICULAR), pp. 6–15, 2014.

[9] K.D. Singh, P. Rawat and J. Bonnin. Cognitive radio for vehicular ad hoc networks (CR-VANETs): approaches and challenges. EURASIP Journal on Wireless Communications and Networking, 2014(1): 1–22, 2014.

[10] M.A. Razzaque, A. Salehi and S.M. Cheraghi. Security and privacy in vehicular ad-hoc networks: survey and the road ahead. In Wireless Networks and Security, pp. 107–132. Springer, 2013.

[11] J.T. Isaac, S. Zeadally and J.S. Cámara. Security attacks and solutions for vehicular ad hoc networks. IET Communications, 4(7): 894–903, 2010.

[12] L. Ertaul and S. Mullapudi. The security problems of vehicular ad hoc networks (vanets) and proposed solutions in securing their operations. In The 2009 International Conference on Wireless Networks (ICWN), pp. 3–9. Citeseer, 2009.

[13] A. Wasef, R. Lu, X. Lin and X. Shen. Complementing public key infrastructure to secure vehicular ad hoc networks [security and privacy in emerging wireless networks]. IEEE Wireless Communications, 17(5): 22–28, 2010.

[14] R.G. Engoulou, M. Bellaïche, S. Pierre and A. Quintero. Vanet security surveys. Computer Communications, 44: 1–13, 2014.

[15] Y. Kim and I. Kim. Security issues in vehicular networks. In 2013 International Conference on Information Networking (ICOIN), pp. 468–472. IEEE, 2013.

[16] M. Al-Rabayah and R. Malaney. A new scalable hybrid routing protocol for vanets. IEEE Transactions on Vehicular Technology, 61(6): 2625–2635, 2012.

[17] J.M. Fuentes, A.I. González-Tablas and A. Ribagorda. Overview of security issues in vehicular ad-hoc networks. 2010.

[18] Y. Qian and N. Moayeri. Design of secure and application-oriented vanets. In Vehicular Technology Conference, 2008. VTC Spring 2008. IEEE, pp. 2794–2799. IEEE, 2008.

[19] H. Hasbullah, I. Ahmed Soomro and J. Ab Manan. Denial of service (dos) attack and its possible solutions in vanet. World Academy of Science, Engineering and Technology (WASET), 65: 411–415, 2010.

[20] J.J. Blum, A. Neiswender and A. Eskandarian. Denial of service attacks on inter-vehicle communication networks. In 11th International IEEE Conference on Intelligent Transportation Systems, 2008, pp. 797–802. IEEE, 2008.

[21] N. Lyamin, A.V. Vinel, M. Jonsson and J. Loo. Real-time detection of denial-of-service attacks in IEEE 802.11 p vehicular networks. IEEE Communications Letters, 18(1): 110–113, 2014.

[22] K. Verma and H. Hasbullah. Ip-chock (filter)-based detection scheme for denial of service (dos) attacks in vanet. In 2014 International Conference on Computer and Information Sciences (ICCOINS), pp. 1–6. IEEE, 2014.

[23] J.R. Douceur. The sybil attack. In Peer-to-peer Systems, pp. 251–260. Springer, 2002.

[24] B. Mishra, S.K. Panigrahy, T.C. Tripathy, D. Jena and S.K. Jena. A secure and efficient message authentication protocol for vanets with privacy preservation. In 2011 World Congress on Information and Communication Technologies (WICT), pp. 880–885. IEEE, 2011.

[25] S. Khan and A.K. Pathan. Wireless Networks and Security: Issues, Challenges and Research Trends. Springer Science & Business Media, 2013.

[26] N.-W. Lo and H.-C. Tsai. Illusion attack on vanet applications-a message plausibility problem. In Globecom Workshops, 2007 IEEE, pp. 1–8. IEEE, 2007.

[27] F. Doetzer, F. Kohlmayer, T. Kosch and M. Strassberger. Secure communication for intersection assistance. In The 2nd International Workshop on Intelligent Transportation, 2005.

[28] R. Panayappan, J.M. Trivedi, A. Studer and A. Perrig. Vanet-based approach for parking space availability. In The fourth ACM International Workshop on Vehicular Ad Hoc Networks, pp. 75–76. ACM, 2007.

[29] H. Sedjelmaci, T. Bouali and S.M. Senouci. Detection and prevention from misbehaving intruders in vehicular networks. In 2014 IEEE Global Communications Conference (GLOBE-COM), pp. 39–44. IEEE, 2014.

[30] R. Chen and J.-M. Park. Ensuring trustworthy spectrum sensing in cognitive radio networks. In IEEE Workshop on Networking Technologies for Software Defined Radio Networks, pp. 110–119, Sept 2006.

[31] Y. Pei, Y.-C. Liang, L. Zhang, K.C. Teh and K.H. Li. Secure communication over miso cognitive radio channels. IEEE Transactions on Wireless Communications, 9(4): 1494–1502, April 2010.

[32] R. Chen, J.-M. Park and J.H. Reed. Defense against primary user emulation attacks in cognitive radio networks. IEEE Journal on Selected Areas in Communications, 26(1): 25–37, Jan 2008.

[33] L. Huang, L. Xie, H. Yu, W. Wang and Y. Yao. Anti-pue attack based on joint position verification in cognitive radio networks. In International Conference on Communications and Mobile Computing (CMC), 2: 169–173, April 2010.

[34] Z. Jin, S. Anand and K.P. Subbalakshmi. Detecting primary user emulation attacks in dynamic spectrum access networks. In IEEE International Conference on Communications (ICC), pp. 1–5, June 2009.

[35] Z. Jin, S. Anand and K.P. Subbalakshmi. Mitigating primary user emulation attacks in dynamic spectrum access networks using hypothesis testing. SIGMOBILE Mobile Computing and Communications Review, 13(2): 74–85, September 2009.

[36] Z. Chen, T. Cooklev, C. Chen and C. Pomalaza-Raez. Modeling primary user emulation attacks and defenses in cognitive radio networks. In The 28th International Performance Computing and Communications Conference (IPCCC), pp. 208–215, Dec 2009.

[37] O. Ureten and N. Serinken. Wireless security through rf fingerprinting. Canadian Journal of Electrical and Computer Engineering, 32(1): 27–33, Winter 2007.

[38] O.R. Afolabi, K. Kim and A. Ahmad. On secure spectrum sensing in cognitive radio networks using emitters electromagnetic signature. In The 18th International Conference on Computer Communications and Networks (ICCCN), pp. 1–5, Aug 2009.

[39] T.C. Clancy and N. Goergen. Security in cognitive radio networks: Threats and mitigation. In The 3rd International Conference on Cognitive Radio Oriented Wireless Networks and Communications, pp. 1–8, May 2008.

[40] O. León, J. Hernández-Serrano and M. Soriano. Securing cognitive radio networks. International Journal of Communication Systems, 23(5): 633–652, 2010.

[41] Q. Pei, H. Li, J. Ma and K. Fan. Defense against objective function attacks in cognitive radio networks. Chinese Journal of Electronics, 20(4): 138–142, 2011.

[42] G. Baldini, T. Sturman, A.R. Biswas, R. Leschhorn, G. Godor and M. Street. Security aspects in software defined radio and cognitive radio networks: A survey and a way ahead. IEEE Communications Surveys Tutorials, 14(2): 355–379, Second 2012.

[43] K. Sakaguchi, F. Chih, T.D. DOAN, M. Togooch, J. Takada and K. Araki. Acu and rsm based radio spectrum management for realization of flexible software defined radio world. IEICE Transactions on Communications, 86(12): 3417–3424, 2003.

[44] A. Brawerman and J.A. Copeland. An anti-cloning framework for software defined radio mobile devices. In International Conference on Communications, 5: 3434–3438, May 2005.

[45] H. Uchikawa, K. Umebayashi and R. Kohn. Secure download system based on software defined radio composed of fpgas. In The 13th IEEE International Symposium on Personal, Indoor and Mobile Radio Communications, 1: 437–441, Sept 2002.

[46] A. Brawerman, D. Blough and B. Bing. Securing the download of radio configuration files for software defined radio devices. In The Second International Workshop on Mobility Management & Wireless Access Protocols, MobiWac '04, pp. 98–105, New York, NY, USA, 2004. ACM.

[47] C. Li, A. Raghunathan and N.K. Jha. An architecture for secure software defined radio. In Design, Automation Test in Europe Conference Exhibition (DATE), pp. 448–453, April 2009.

[48] D. Cabric, S.M. Mishra and R.W. Brodersen. Implementation issues in spectrum sensing for cognitive radios. In The Thirty-Eighth Asilomar Conference on Signals, Systems and Computers, 1: 772–776, Nov 2004.

[49] C. Mathur and K. Subbalakshmi. Security issues in cognitive radio networks. In Cognitive Networks: Towards Self-aware Networks, Wiley, New York, pp. 284–293, 2007.

[50] R. Chen, J.-M. Park and K. Bian. Robust distributed spectrum sensing in cognitive radio networks. In The 27th Conference on Computer Communications (INFOCOM), pp. 31–35, April 2008.

[51] F. Zhu and S.-W. Seo. Enhanced robust cooperative spectrum sensing in cognitive radio. Journal of Communications and Networks, 11(2): 122–133, April 2009.

[52] W. Wang, H. Li, Y.L. Sun and Z. Han. Attack-proof collaborative spectrum sensing in cognitive radio networks. In The 43rd Annual Conference on Information Sciences and Systems (CISS), pp. 130–134, March 2009.

[53] A.S. Rawat, P. Anand, H. Chen and P.K. Varshney. Countering byzantine attacks in cognitive radio networks. In International Conference on Acoustics Speech and Signal Processing (ICASSP), pp. 3098–3101, March 2010.

[54] K. Bian and J.-M. Park. Mac-layer misbehaviors in multi-hop cognitive radio networks.

[55] Ieee-802.22 working group on wireless regional area network.

[56] L. Ma, C.-C. Shen and B. Ryu. Single-radio adaptive channel algorithm for spectrum agile wireless ad hoc networks. In IEEE International Symposium on New Frontiers in Dynamic Spectrum Access Networks (DySPAN), pp. 547–558, April 2007.

[57] L. Lazos, S. Liu and M. Krunz. Mitigating control-channel jamming attacks in multi-channel ad hoc networks. In The Second ACM Conference on Wireless Network Security, WiSec '09, pp. 169–180, New York, NY, USA, 2009. ACM.

[58] A. Chan, X. Liu, G. Noubir and B. Thapa. Broadcast control channel jamming: Resilience and identification of traitors. In IEEE International Symposium on Information Theory, pp. 2496–2500, June 2007.

[59] P. Tague, M. Li and R. Poovendran. Mitigation of control channel jamming under node capture attacks. IEEE Transactions on Mobile Computing, 8(9): 1221–1234, Sept 2009.

[60] G.A. Safdar and M. O'Neill. Common control channel security framework for cognitive radio networks. In Vehicular Technology Conference (VTC), pp. 1–5, April 2009.

9

Vehicular Clouds Based on GroupConnect and Self-Organization

Marat Zhanikeev

ABSTRACT

GroupConnect is a recent wireless technology that has all the features of mobile clouds but also pools resources of an entire group into a single virtual wireless node that can enjoy boosted throughput and otherwise optimize resources both inside the group and on the in- and out-group border. GroupConnect can be applied to vehicular networks in the same way as it has been applied to human wireless networks before, where the largest distinction between the two lies in mobility patterns. This chapter talks specifically about vehicular clouds which are defined as cloud services running on top of networks of vehicles. This chapter shows that GroupConnect is an essential guarantee that cloud services are feasible in practice. It is also shown that the presence of the global supervisor in form of a cloud service provider simplifies self-organization of vehicles. The two specific cloud services discussed in this chapter are virtual storage and sensor networks.

Vehicular clouds (not just networks) are actively discussed in research. The common terms are *Vehicular Cloud Computing* (VCC) and *Internet of Vehicles* (IoV), both borrowing from similar terms that have been previously defined in non-vehicle areas. The ultimate goal of vehicular clouds is to implement *smart ITS* with enhanced decision-making and improved utilization of resources hidden in various fleets of vehicles. When discussing *fleets* specifically, the subject of *smart cities* [3]

Department of Artificial Intelligence, Computer Science and Systems Engineering, Kyushu Institute of Technology, Fukuoka Prefecture, Japan.
Email: maratishe@gmail.com

and *smart grid* are closely related—the former describes a city where vehicle fleet are an intrinsic part of its infrastructure while the latter is a specific extension to the future generation of power grids.

For a large share of academic literature, the cloud function is an extension to the networking function of Vehicular Ad hoc NETworks (VANETs). See a survey that traces this evolution in [10]. It is common for such research to take roots in VANETs even when clouds play a major role in orchestrating interactions [11].

However, there is also literature that considers clouds of vehicles its own kind of academic discipline. The main justification in such literature is that, unlike VANETs, vehicles in clouds are normally connected to the cloud on the individual basis. In other words, the core problem to VANETs (and the earlier and more general problem of MANETs) of end-to-end routing in loosely coupled networks of vehicles is not there when the vehicles are part of a cloud service. An example of such research is the subject of *Internet of Vehicles* (IoV) [13] which focuses on the various usecases of connected vehicles where the VANET-like usecase of isolated networking can be viewed as a side function. Note that the term *connected vehicles* itself is also common [16] and refers to vehicles that get continuous access to 3G/4G wireless networks. This chapter discusses this very part in detail, offering two specific models for how connected vehicles can be implemented in practice. When security is property implemented, connected vehicles can allow access to the CAN bus, thus opening up the information flow all the way from vehicle internals to back-ends in clouds [14].

Based on the above, the term *vehicular clouds* in this chapter refers to a design in which vehicles are connected to the Internet on individual basis on one end, which makes it possible for cloud services to incorporate individual vehicles into local copies of themselves (offload, distribution). The biggest change of paradigm here is that cloud services become fleet managers. This crude definition is enhanced in this chapter with the specific subjects of fleet creation and self-organization—specifically, the grouping of individual members of the larger fleet.

This chapter follows the obvious relation of vehicular clouds to the subject of *mobile clouds* [22], where, just as vehicular clouds are viewed as extension to VANETs, mobile clouds are viewed as extension to MANETs. The specific part shared in this chapter is the clean separation of *local* vs. *remote* networking and the need to balance between the two. Note that the topic of mobile clouds has existed for several years and already has many existing platforms [26]. Also note that this connection is not considered by majority of existing literature on vehicular clouds [10].

As a form of connection to mobile clouds, the key component in this chapter is *GroupConnect* [1]. The technology is rival to mobile clouds with several unique features like *MultiConnect* and *virtualization of resources*. See [4] for early work and [1] for more details on these features. A practical usecase for GroupConnect in form of a new generation of a converged wireless university campus can be found in [5]. This chapter makes the case that GroupConnect can help implement vehicular clouds with very few alterations, mostly in areas of mobility patterns, power efficiency, etc.

This chapter is centered around practical cloud services running on top of fleets of vehicles. A large-scale infrastructure is assumed—see [3] for an example of a large-scale infrastructure built about an *EV battery replacement* service. This chapter shows that modeling for vehicular clouds needs to be more complex than was earlier presented in [3]. The model needs geographical context which can be provided by real traces—this chapter discusses *crawdad* [42] and *maps2graphs* [9] traces. However, the model also has advanced essential components like distribution density, context for local grouping, etc., some of which are discussed but not implemented. It should be stressed that such a model is not found in existing literature and can be considered as the first attempt for formulate a modeling framework for vehicular clouds. The specific model in this chapter describes a fleet of up to 800 k vehicles in northern Kyushu, Japan.

This chapter has the following structure. Section 1 establishes all the core assumptions about vehicular networks, connected vehicles, and, ultimately, vehicular clouds. Section 2 explains GroupConnect and self-organization in context of vehicular clouds. Section 3 explains the relation between vehicular and mobile clouds and enhances the description of the latter. Section 4 formulates the model, practical cloud services, and simulation design and performs the related analysis. Section 5 summarizes the chapter.

1. Core Assumptions about Vehicular Clouds

This chapter provides all the necessary background for vehicular clouds including several technologies starting from 3G/4G/5G wireless networks, populations/fleets in vehicular networks and clouds, etc. Note that some of these subjects are not related and come together for the first time in vehicular clouds. For example, this strongly applies to the concept of *population* (fleets for vehicles) where the two similar concepts in clouds and vehicular network are discussed in a shared context for the first time in this chapter. For evolution of wireless networks, good overviews can be found in [32], [33] and [34]. This subject is related in the area of grouping and self-organization —this problem is part of current research on 4G/5G networks and is also part of the technology considered in this chapter.

The concept of *connected vehicles* is paid special attention in this section because it is a unique feature for vehicular clouds and is what makes them distinct from VANETs—that said, note that a large part of existing literature views vehicular clouds as natural extension to VANETs.

Populations/fleets and their management is found in two separate areas. In vehicles, it is about fleets of vehicles, as in the example of the EV fleet brought together by the battery replacement service [3]. In clouds, distributed resource which consists of a large number of individual items (virtual machines, storage spaces, etc.) is also viewed as populations that require intensive attention when building and maintaining [2]. The two populations are brought together for the first time in this section.

1.1 4G and 5G Wireless Networks

Today's cellular (or wireless in a broader sense) networks are in the pre-4G state. It is also common to refer to them as 3.8 G networks—the term often used to describe the Long-Term Evolution (LTE) networks widespread today. The main problem with cellular networks todays is high latency and low capacity. For example, it has been measured that a common cellular subscriber today gets the average throughput of 400 kbps [4]. Other measurements in the same study showed that one can get higher throughput but not by switching to LTE. Instead, it was found that higher throughput can be attributed to younger networks (which can be 3G), specific times of day or days of week. In other words, capacity in today's cellular networks is highly affected by congestion.

While an average subscriber can live with low throughput, other studies show that this situation makes professional use nearly impossible—the study in [5] makes the case for educational use of existing 3G/LTE networks. Note that the GroupConnect technology was originally viewed as a means of resolving this problem in practice, where [5] shows that it can help students boost their throughput within a university campus.

4G networks are viewed as further evolution of the LTE technology [32]. The new technology is called LTE-Advanced (LTE-A) and is actively discussed in literature today [33]. It has many components which are useful in practice today. Let us consider a small subset of them, which closely relates to the GroupConnect and its application to vehicular clouds. Their detailed descriptions can be found in [32][33].

Microcells (other names are eNBs, femtocells, picocells, etc.) is a technology that fulfills two objectives: it distributes cellular service via a large number of smaller local cellular stations and, by doing this, allows the network to offload some of its load to the newly created network of microcells. Microcells effectively change the topology on the global scale because base station are now one level higher in the topology tree while microcells are put in charge of talking to individual subscribers. This change makes it possible to support much higher rates because microcells can maintain stable high-rate connections to base station continuously while current network requires the subscriber to negotiate a time slot for every connection. Some of the efficiency comes from the fact that 4G networks plan to switch to frequency multiplex (OFDM) rather than time-based access slots today.

MIMO (Multiple Input, Multiple Output) solves the subscriber side of the problem with high-rate communication. Subscriber's device can have multiple software-controlled antenna which can be pooled on demand to boost capacity of a given connection. Both microcells and subscribers are expected to implement the technology. In fact, MIMO is already part of many WiFi routers and smartphones today.

4G/5G networks will also depend on *device-to-device* (d2d) and *machine-to-machine* (m2m) modes of operation. The difference between the two is only in the type of the device—smartphones and other wireless terminals are called *devices* while any other network-aware *things* in the Internet of Things (IoT) are referred to as *machines*. Naming difference aside, a the technologies are the same and assume that two devices can talks directly to each other.

It is important to unfold the term *directly* here. In 4G/5G, direct does not mean *direct between two devices*—although some current research considers such usecases as well—but rather refers to connections between two devices that use existing 4G/5G infrastructure as intermediate routers. For example, a device may request a microcell to route its packets to another device within the same cell. Even when the cell is shared, current 4G/5G standards do not consider cases when devices talk directly to each other. Note that GroupConnect, on the other hand, *always* assumes that two devices connect to each other without any mediating infrastructure.

The above newly introduced technologies can support a number of new behaviors. Existing research discusses *cognitive, cooperative,* and *opportunistic* networking in the context of 4G/5G infrastructure [35]. There are also new problems, where the problem of interference from the higher density of local radio communication has already been recognized [34]. This general discussion is part of the research on 5G networks which discusses all the above newly introduced subjects [34]. The specific technologies leading up and helping build 5G today are discussed in [33].

Even with the substantially enhanced subject of future wireless networks above, there are missing pieces. For example, the subject of *MultiConnect*—ability to support two distinct physical types of connections (WiFi and 3G, for a trivial example)—is almost entirely missing in current discussions [4]. However, earlier work in [1] and this chapter shows that MultiConnect is crucial for feasibility in group communications. Vehicular grouping has to be discussed in the frame of distinct mobility patterns which can be either help or hurt performance of such groups, depending on specific situations.

Local wireless network coordination (positioning is another term found in literature) is also missing from 4G/5G literature but exists as a separate research subject that goes through its active phase today [39]. Note that GroupConnect and network coordination are closely related subjects simply because efficient coordination can help form robust groups which, in turn, can improve performance of GroupConnect.

1.2 Connected Vehicles

Figure 1 is a visual guide into the subject of *connected vehicles*. The visual first shows all the major components in the center and then shows how the components can help build two distinct technologies. Note that although the two combinations may appear similar at the first sight, they offer very different utilities when it comes to vehicular clouds, that is when individual vehicles have to be utilized as part of a cloud service running on top of a fleet of vehicles. The basic components are:

- *CAN bus* and otherwise in-car networking;
- person and smartphone;
- 3G/LTE connection to the Internet.

Let us consider the two distinct combinations of these components, in order of first the existing *conventional* and then the newly introduced *advanced* cases.

Conventional technology assumes that vehicle owner uses his/her own smartphone to connect the car to the cloud. It has already been shown that it is possible to use the smartphone as a medium between the CAN bus inside the vehicle and the cloud. Clearly, security is extremely important for such mediation [14]. Such a technology has the obvious consumer appeal as well—owners can view the status of their cars via a simple smartphone app. Since it is not difficult to implement such a technology, many products are already available today [16][18][19][20].

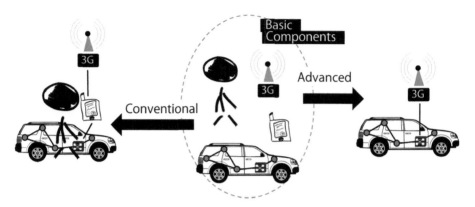

Figure 1. The two conceptual foundations for vehicular networks.

Advanced technology makes one small adjustment to the conventional case—cars here get their own smartphones rather than wait for their owners to bring them when taking a ride. Note that adding a smartphone to the list of built-in hardware has negligible effect of power consumption, even for Electric Vehicles (EVs) [3]. Security issues here are the same as for the conventional case but restrictions have to be rougher simply because your vehicle gets to keep its smartphone even when you get out and turn it off. It is understandable that early versions of this technology will shy away from providing too deep an access into CAN messaging. However, most cloud services would be satisfied with read-only access (temperature, GPS, etc.), so read-only access can be hard-wired into the physical devices that provides such access. A prototype of the technology described in this chapter is currently developed by this author in form of a simple car navigator.

The obvious advantage of the advanced technology is that it can run in the *always on* mode, supporting a continuous connection to the 3G/LTE network. However, this is the very feature that makes such vehicles attractive to cloud services which can now rely on vehicles to be connected at all times rather than getting access to cars intermittently. It can be argued that cloud service providers would probably reject vehicles that do not provide a 24-hour connectivity. This chapter shows further on that even with always-on connected cars, there are other metrics which makes grouping specifically and fleet management in general a non-trivial technology.

Let it take a forward peek into cloud services which can be built on top of the advanced technology above. If multiple vehicles are parked together on a regular basic, than the cloud service can include them as part of its storage service, where hard disks installed in cars would be rented out to the cloud service as virtual storage space. When in motion, on top of vehicle-to-vehicle (v2v) connections which are already possible today [13], a cloud service can manage the fleet in such a way that it can collect sensor data for a given geographical area at a given time of day. Note that for both these cases, throughput boosting via GroupConnect is essential. Also note that while GroupConnect remains fully operational when there is only one vehicle (it becomes the ordinary SingleConnect), it becomes increasingly more useful when more vehicles are added to the group.

1.3 Vehicle Fleets and Cloud Populations

The subject of grouping and its role in GroupConnect has already been mentioned several times in this chapter. This section shows that vehicular clouds also need *population/fleet management* layer on top of the fundamental layer of unit grouping.

The subject itself of grouping is not new in research and even exists beyond GroupConnect. For example, recent methods in MANETs discovered that end-to-end routing can be made more efficient if nodes are grouped first and routing is performed between the groups first and then down to individual nodes. In MANET research, such methods are referred to as *group* or *flock* routing [36][37]. Note that such groups closely resemble mobile clouds and GroupConnect which also make a clear distinction between local and remote (often referred to as in- versus out-group) traffic. The similarity between MANETs and GroupConnect, however, ends at this single feature. Specifically, e2e routing is absent in mobile clouds and GroupConnect, which both assume that each device has a direct connection to the Internet.

Alternative kinds of grouping can be found in Delay-Tolerant Networks (DTNs) [40]. The distinction between DTNs and MANETs is mostly in the area of grouping where DTNs initially assumes that even grouping is ad hoc while MANETs assume groups to be fairly stable (if present) while routes between individual nodes are ad hoc. DTN is revisited several times throughout this chapter because it has direct relation to GroupConnect.

The current subject of VANETs [15] has a grouping component to it, which derives mostly from the difference in mobility patterns between humans and cars (the latter travel on roads and have a natural tendency towards grouping). Finally, mobile clouds assume that grouping is part of their natural operation [22]—the feature it shares with GroupConnect [1].

Having reviewed research on grouping, it is important to distinguish grouping from population and/or fleet management. There are many examples of fleets of vehicles in practice—see the example of EV fleet with battery replacement in [3]. Just as there are many existing populations in clouds—the example of a video streaming serving which improves streaming quality by maintaining and dynamically improving a population of video sources in the cloud [2]. While these two populations come from very different areas, they come together in the subject of vehicular clouds.

Let us walk the process of creating a given vehicular cloud. First, we need a fleet of vehicles that have hardware that can host software that communicates to the core software of the service running in the cloud. This is a long but essential chain of elements. The *advanced technology* of connected vehicles above shows that this feature can be implemented in vehicles today. The next step is to make the fleet known to the cloud service in question. This can be done via a *cloud service provider* discussed further in this section. It is convenient in practice to implement both cloud service provider and cloud service in the same software—meaning that the cloud service is now directly involved with managing the fleet of vehicles as well.

The terms *population* and *fleet* are merged at this point. If the cloud service becomes the manager of a fleet of vehicles, then the fleet effectively becomes an extension of cloud infrastructure (hardware, data centers, etc.), which further leads to the conclusion that a portion of the population of resources hosted on the main cloud can now be transferred to this new extension and continue to run in vehicles.

Coming back to the comparison between grouping and population management, the difference in the concepts comes to the difference in layers of management. While grouping of vehicles can be implement on small scale, the groups are useless unless used as part of a larger fleet of vehicles. The terms *fleet* and *population* are used interchangeably in this chapter.

Note that the literature on population/fleet management is extremely scarce. For example, survey in [10] comes very close to defining the term but does not go beyond defining several *aaS cloud services, most of which depend on or benefit from fleet management. A perfect scenario for fleet management can be found in [11] but the paper itself does not formulate the concept, instead satisfied with a specific small-scale fleet participating in a disaster relief operation.

Here are some examples of useful fleets:

- EV battery replacement service provider, where the fleet contains all the vehicles that use such a service [3];
- fleets of cars from the same maker or for each specific model—take the recent example of the Mirai fuel-cell car produced by Toyota;
- cars in which owners purchase and install the same model of car navigator—this is the case which can support a simple prototype of the cloud service discussed in this chapter provided that navigator has built-it hard-disk and 3G connectivity.

2. Group Connect and Self-Organization

As was mentioned above, GroupConnect is a kind of mobile cloud. Good surveys on mobile clouds can be found at [21] and [22]. This section shows that GroupConnect has more features than are found in mobile clouds, specifically the MultiConnect and resource virtualization, both of which play crucial role in vehicular clouds. This section also shows that GroupConnect does restrict the degree of freedom of self-organization, but at the same time is helps create robust local groups which are more attractive when viewed by cloud services. While this section is not specific to vehicular clouds, all the technologies are discussed in the context of running on top of vehicular fleets.

2.1 MultiConnect and GroupConnect Basics

Figure 2 shows the three fundamental wireless technologies in context of vehicular clouds. The same comparison is possible for human wireless networks in which case cars would be replaced by humans and VANETs by MANETs. The rest of this section reviews each communication mode in detail.

SingleConnect is when a vehicle can use only one wireless connectivity at a time. The figure specifically shows the case of 3G versus WiFi connections, in which case most operating systems (including those in smartphones) would select WiFi as the default technology and ignore the other. More details on default behavior can be found in [4]. All in all, this is the traditional form of networking.

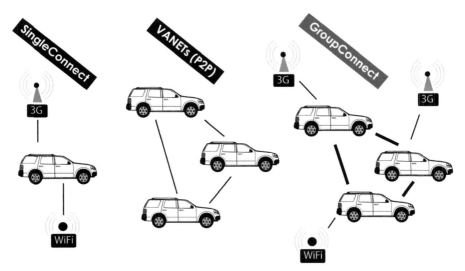

Figure 2. Comparison of GroupConnect to other modes in wireless networks.

VANETs (or MANETs for humans) is a P2P technology [11][15], where vehicles self-organize into networks automatically when the need arises. DTN form is a specific kind of MANETs [40], in which end-to-end routes can exist even when disrupted multiple times. Note that there are many kinds of DTNs including those which optimize for delay-minimized e2e delivery [41]. While retaining the same basic form, MANETs have recently developed some forms of grouping [36][37]. The main problem with MANETs is the absence of the Internet connection. In fact, MANETs are mostly concerned with e2e connectivity within the MANETs themselves and therefore do not consider Internet connectivity as an essential function. When Internet connection is required, it is common in existing research to select one node as *gateway* between the MANET and the outer Internet.

GroupConnect assumes that vehicles can and do support parallel MultiConnect by being able to communicate to 3G network at the time same as exchanging information with another vehicle in the group. Note that only the in-group technology has to be the same for all vehicles, while each individual vehicle is free to use any available technology for out-group communication (to the Internet). The specific combination widely available in smartphones today is 3G/LTE (out-group) with WiFi Direct (in-group) [4]. The device with WiFi Direct can also support WiFi connection in parallel with multiple WiFi Direct sessions to other devices in the group.

GroupConnect is all about *resource virtualization*. All the resources—networking, CPU, storage, etc.—is aggregated into a single node which then supports a single virtual connection to the Internet. This formulation makes perfect practical sense for vehicular clouds because all vehicles use the same exact service and do not mind sharing parts of their own work with others, assuming that other vehicles help itself complete its own work faster. This, in a nutshell, is what GroupConnect is about, from the practical viewpoint.

GroupConnect has the following distinctions. In-group communication can support much higher throughput than out-group communication by any individual vehicle. The study in [4] shows that WiFi Direct can support up to 30 Mbps rates at 10 m distance. By extension, this also means that the aggregate of all out-group rates experienced by individual vehicles is not expected to exceed the in-group rate because this would create surplus in-group capacity which cannot be filled with additional out-group traffic.

Note the nature of in-group communication supported by WiFi Direct today makes GroupConnect somewhat similar to DTN [7]. WiFi Direct connections are not continuous like in the traditional WiFi but are rather short-term sessions intended for exchanges of bulk between nodes rather than continuous communication in both directions. Further in this chapter it is shown that this is not a problem for cloud services which, in fact, follow the same rules in own design.

It is a good place to stress on the connection between mobile clouds and GroupConnect. Mobile clouds use a similar concept of in- and out-group communication [23], where the cloud service itself runs locally but needs to sync its work with the master copy in the core cloud. However, mobile clouds do not assume MultiConnect and cannot formulate the problem as virtualization of resources, with subsequent optimization of in- and out-group traffic bulks. This discussion is expanded further in this chapter.

2.2 Clouds and Throughput Boosting

When GroupConnect is presented by itself, it is easy to miss the advantages in one of its popular operation modes—*throughput boosting*. Note the virtualizations which underlies the technology offers many other usecases, some with diagonally opposing properties. To clarify the throughput boosting feature of GroupConnect, this section discusses in a lineup of other wireless technologies from 3G to future 5G.

Figure 3 is based on earlier studies [4] and overviews in [1] and [34]. The following technologies are presented in increasing capacity left to right:

- 3G/LTE can support up to 1 Mbps in practice [4]—we intentionally ignore the nominal rates supported by the technologies;
- most cloud APIs today support between 2 and 3 Mbps, with occasional outliers—the limit is more a product of traffic shaping applied at cloud side rather than the limitation of the e2e throughput which the network can support;
- WiFi (normal operation) can support much higher traffic rates than those supported by most cloud APIs;
- WiFi Direct is known to support practical rates higher than those by WiFi [4], in the same device, same distance, etc.—this artifact is due to the fact that WiFi Direct is a one-hop network while connections to WiFi routers are normally part of a longer path;
- theoretically, 4G/5G microcells can support up to 1 Gbps rates [34].

The figure has two extra notations. One shows that cloud APIs are a limiting factor on both 5G (when they arrive) and WiFi connections. This means that even if WiFi direct or LTE connections provided higher rates, they would be wasted when

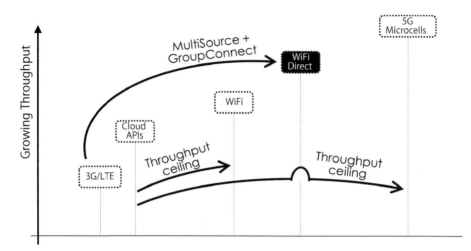

Figure 3. Distinct wireless technologies in context of clouds, cloud APIs and throughput boosting.

used in the traditional SingleConnect design. The other notation show that multiple 3G/LTE connections can be pooled up to the aggregate rate supported by WiFi Direct, provided the design implements GroupConnect and MultiSource.

MultiSource is newly introduced here and has to be properly defined. MultiSource technology makes it possible to transfer the same content between two sides via multiple routes/connections/connectivities. A good example of a MultiSource technology is the video streaming in clouds method introduced in [2]. It has been shown to work in browsers using Web-Sockets (HTML5) [6] where it can support up to slightly less than 1 Gbps rates. Next subsection expands on the MultiSource technology in the context of vehicular clouds.

Throughput boosting is essential to some forms of vehicular clouds, and can improve performance even if it is not strictly essential. Some justification for this statement can be found in mobile clouds today where balancing between in- and out-group traffic is difficult also because content comes in crude bulks of individual files [22]. MultiSource can help by splitting files into chunks and sending chunks over different connections. MultiSource is only distantly related to the MultiPath TCP protocol.

2.3 Example of a Human GroupConnect

This section presents an example of human GroupConnect which also shows how MultiSource can be implemented and used in practice. Since this example also uses clouds as back-end, it can be applied to vehicular clouds without changes. The example comes from a larger work in [1].

The problem is as follows. Students at university campuses today heavily rely on 3G/LTE or local WiFi networks. With the rate-shaping on the part of cloud APIs (see previous subsection), in addition to rate limitations of the two wireless technologies themselves, educational use of such networks is limited to unacceptably small content size. The objective is to improve this problem using a combination of GroupConnect and MultiSource technologies.

Let us walk the process of creating a new wireless campus in Figure 4:

1. University makes an application (app) for smartphones that have WiFi Direct and can therefore support MultiConnect. The GroupConnect functionality itself is implemented in user space by the application.
2. Students get the app in the most common way—from University COOP. The app ships with a special security code that only allows for wireless meetings for other such apps. This is a useful feature (not a bug) that easily limits the physical area where the application can be applied. In this specific case, it can only be used when two or more students from the same university come in proximity to each other.
3. Students *meet* wirelessly. In this simple implementation, groups are built by randomly selecting nearby users (where there are too many), but advanced versions can include local coordination [39]. When the handshake is successful, one student can delegate parts of its own tasks to other students, or vice versa.
4. Finally, MultiSource is implemented by sharing *tokens*. Most cloud APIs allow access via tokens, some of which are generated online for specific sessions (file downloads, etc.). The tokens can be shared with other users in the group and used to download, for example, other sections of the same file. By compensating for mobility with a certain level of delegation redundancy, it is possible to complete the entire task in a fraction of time that it would take in the SingleConnect mode.

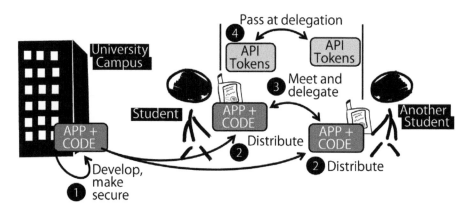

Figure 4. Operation of software that implements GroupConnect in a wireless university campus.

Repeating an earlier statement, while the above example is based on human wireless interactions, the basic concept of MultiSource is applicable to vehicular clouds as well. In fact, since vehicular fleets are managed by cloud services, it is likely that clouds themselves will implement MultiSource in each vehicle in an effort to boost throughput and, therefore, the volume of content which can be exchanged between a group of vehicles and the cloud. In view of the two practical technologies discussed further in this section, the MultiSource component is essential.

3. From Mobile to Vehicular Clouds

The subject of *mobile clouds* is well-known, with good overviews in [21][22]. GroupConnect is a related technology but has a larger scope [4]. The specific GroupConnect technology for vehicular clouds yet has unique properties while sharing the same basic features with the generic GroupConnect. This section first discusses the core problem resolved by mobile clouds and then proceeds to vehicular clouds and the newly posed problems, such as fleet generation and management. Another major issue discussed in this section is the challenge of modeling of large-scale vehicular clouds. This section shows that, while several components can be borrowed from related models, no one model can cover the full range of elements required for realistic modeling of vehicular clouds.

3.1 The Mobile Cloud Problem

At the core of mobile clouds is the *resource optimization* problem [22]. Most mobile cloud technologies clearly divide traffic into local versus remote parts [23]. The other common feature is energy-efficiency where mobile devices attempt to lower battery consumption [21][22].

The problem of battery capacity is especially acute in mobile clouds because cloud services require nodes to stay in active state for extended periods of time. There is research that looked into the tradeoff between the size of the battery and its utility for mobile clouds [25]. Note that this whole issue does not exist in vehicular clouds because car batteries are huge compared to any battery available to smartphone-level devices.

On a distant relation to the battery efficiency problem, some mobile cloud research discusses the game theory involved into incentive-based social participation of mobile subscribers in a mobile cloud [26]. In majority of cases, the incentive exists to offset the loss of battery power due to extended participation. This subject has some relation to vehicular clouds as well. However, it should be noted that battery power (consumption) costs very little in vehicular clouds, again, because car batteries are relatively large. However, some payback for the cloud resource (storage, sensor data, etc.) should be expected by car owners when they install and enable a given cloud service in their cars. This is an interesting subject but cannot be accommodated in this chapter. It will be revisited in future publications.

As another viewpoint at the technology, mobile clouds are often considered as a case of *computational offload*. This fits into the in/out-group framework—local traffic is considered as an offload of having to perform the same set of actions in the core cloud. This formulation is not perfectly accurate because it ignores the traffic specifically and other efforts in general that are required to sync the states between local and remote parts of the mobile cloud. However, mobile clouds are efficient as long as the cost of syncing is lower than the cost of running computation in the core cloud. One of the cloud services analyzed further in this section is also a kind of offload technology.

Mobile clouds are spreading fast. There are many existing platforms for a wide range of practical targets:

- a generic/multipurpose platform for computation offload to mobile clouds [26];
- thin clients running and migrating in local mobile clouds [24];
- offload and remote execution of tasks on mobile clouds [27];
- offload that balances performance and energy consumption [28];
- parallel processing and optimization specific to mobile clouds [29];
- BigData/MapReduce platform specific to mobile clouds [30];
- a software platform that extends traditional MANETs to mobile clouds by, essentially, running cloud services on MANET topologies [31].

In addition to the points expressed earlier in this chapter, the following are the major distinctions and similarities between GroupConnect and mobile clouds:

- in mobile clouds nodes are independent with a fixed master node used for syncing between local and remote, while in GroupConnect there is no clear or fixed syncing node, all nodes have equal shares in syncing activity;
- optimization of resources in mobile clouds is done for individual nodes first and then for the whole mobile cloud, while GroupConnect virtualizes resources for the entire group and therefore optimization is always/only global;
- the only similarity between the two technologies is in clear separation between local and remote traffic.

Self-organization is a separate issue considered further in this section. Note that this issue is not considered in mobile clouds because such networks are normally fixed in topology and have a clear *gateway* (master node for syncing). However, there are attempts to introduce self-organization into mobile clouds, a good example of which is the Satin platform [38].

3.2 Vehicular Clouds and Services

Figure 5 shows the full taxonomy of the type of vehicular networks and clearly separates the types which are useful to vehicular clouds from all the others. At the highest tier, cars can be split into *meeting likely* versus *rarely meet*. The entire *rarely meet* subtree is out of scope of vehicular clouds—at least in the definition used in this chapter.

Note that the *rarely meet* subtree has all the traditional mobile networks including MANETs (VANETs) and DTN. This is the realm of opportunistic networking where mobile nodes have to catch any occasional chance they get to communicate to other nodes. As was mentioned earlier in this chapter, this form of communication is not suitable for cloud services which require a certain level of consistency.

This means that only the *meeting likely* subtree can host technologies which are suitable for cloud services. The obvious two forms of likely meetings are when vehicles are *parked together* and when they *travel together*. Note that there can be an overlap between the two because if two cars are parked together, with a certain likelihood they can share a portion of their subsequent travel. Analysis further in this section expands further on this relation.

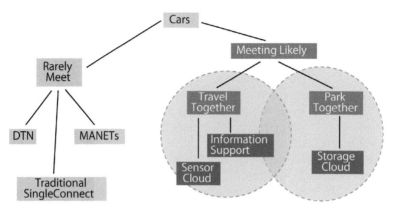

Figure 5. Taxonomy of cloud technologies enabled by GroupConnect.

These two types of wireless meetings can support at least the three obvious cloud services shown in the figure. The rest of this subsection discusses each cloud service in detail.

Sensor Cloud can work with still vehicles but normally requires vehicles to be in motion where they play the role of moving sensors, from the viewpoint of the cloud. A wide range of sensor activity is discussed in literature, from trivial temperature measurements, to detection of traffic jabs or accidents, to recording audio and video for certain locations. Depending on measurement target, vehicles and, ultimately, cloud services, can benefit from the traffic boost provided by GroupConnect.

Information Support is the same basic technology as was explained for Sensor Cloud above, but the direction of data flow is reversed. The popular example considered in research literature is *realtime decision-making* supported by clouds. Clouds here provide the expertise backed by the BigData while vehicles get this expertise in digested form. However, even the digested form may require throughputs which can only be supported by vehicles with boosted throughput, that is, active members of a GroupConnect session, while in motion.

Storage Cloud is the easiest to define. Each vehicle installs a local hard disk and simply rents it as storage space which can be accessed (in both directions) by the respective cloud service. This mode is suitable only for still groups of cars because even low levels of mobility would make content syncing between cars very difficult. However, a normal car is parked for long periods of time every day, where overnight parking is probably the longest period. Cars also naturally group in parking spaces, around or next to tenant houses, etc. This further strengthens this usecase—many still cars over long periods of time is a perfect mobile cloud (traditionally) and vehicular cloud (in this chapter). Cloud service can plan for large bulk transfers between cars as well as between cars and clouds. At the same time, many still cars provide the maximum boosting ratio for throughput which, in turn, enhances syncing capability between groups of vehicles and the cloud.

Analysis further in this chapter focuses on two of the above three services—storage cloud and sensor cloud.

3.3 Cloud Service Providers and Their Fleets

The link between *cloud populations* and *vehicular fleets* was established earlier in this chapter. This subsection further enhances this foundation and shows how clouds can use the fleets to run cloud services. EV battery replacement service is an example of vehicular fleet but not cloud population [3].

Figure 6 shows the map of Kyushu Island of Japan (leftmost), then zooms into the Fukuoka prefecture, which is the point of focus for the example service provider in the figure. Like most cities on the planet, Fukuoka city has a complex network of roads both in immediate proximity to the city and connecting Fukuoka with smaller neighboring cities. For most cloud services, the main advantage of running on top of vehicular clouds is the geographical distribution they can provide. The specific example that fits this description is the *sensor cloud* defined above, which is feasible only if it provides sufficient geographical coverage.

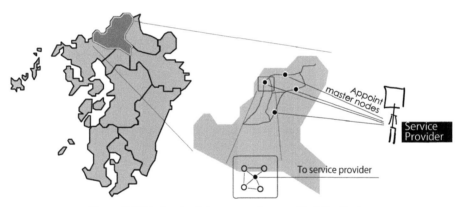

Figure 6. Vehicular clouds in context of geographical distribution.

Geographical distribution is not difficult to achieve in practice, as long as the service provider oversees the entire fleet—in has already been defined as the essential feature of vehicular clouds earlier in this chapter. First, GPS makes spatial positioning easy. When the same GPS location is shared by multiple vehicles, local positioning can be done using one of many available tools [39].

The remaining part is the self-organization part of GroupConnect. In MANETs or VANETs, self-organization assumes that *master* nodes are selected in isolation using a robust algorithm. Such algorithms describe both the procedure for selecting the master node from several candidates and procedure for re-selection when current master disappears or the group cannot support its current configuration without a new master.

With connected vehicles, this form of self-organization can be replaced by trivial assignment. Since service provider has full grasp of all the vehicles that are part of the fleet, the assignment process is easy. The local group around the master node can be formed by the master node itself, where other nodes are notified of its *master* status during or immediately after the handshake. Alternatively, again, based on the notion of connected cars, service provider can choose to form groups *manually* by enforcing grouping links for each vehicle in the fleet. It is obvious, that the local grouping method is much more robust and much less prone to errors due to mobility, GPS discrepancies, etc.

Let us consider two specific cloud services from the viewpoint of the above concept of fleet management.

In *storage clouds*, service provider needs geographical distribution and fairly large persistent groups. There is a tradeoff between the two parameters and, arguably, degree of geographical diversity can be lowered for the sake of having robust groups. Analysis further in this chapter shows that groups decrease in side further away from the city, which means that the tradeoff can be simply the intersection point between the two curves.

In *sensor clouds*, service provider only needs geographical distribution. It can also benefit from larger and more persistent group but this is not a tradeoff because only geographical diversity is essential. Also, even small groups can support the throughput required by the cloud. As was shown earlier in this chapter, 3G/LTE networks today can support around 400 kbps throughputs on average (in practice rather than nominal). Two vehicles can effectively double this number. Sensor data that exceeds 1 Mbps rates is rare in practice. Note that only one set of data has to be transferred for the entire group because all the vehicles are roughly in the same location and data from all vehicles is redundant, from the viewpoint of the cloud service that collects it for online or offline analysis.

3.4 The Challenge of Vehicular Cloud Modeling

This section shows that it is very difficult to model vehicular clouds. The entire model is presented in this section, while its simplified version is used for analysis further in this chapter. Note that the below model is much more complex than the one used for EV battery replacement infrastructure in [3]. The below model borrows some of its components but some others are uniquely defined for vehicular clouds.

A small note has to be made about *traces* and *trace-based simulation*. Vehicular simulations today can be split into two major kinds: detailed but small scale models normally focusing on a single intersection or otherwise a small network of roads [15], and simulations based on real traces which normally come from large-scale environments and/or infrastructures [42]. This section shows that neigher is helpful for simulations of vehicular clouds.

Figure 7 shows two plots drawn from real traces. The left plot shows a randomly selected snapshot from a *crawdad* trace for a small Rome taxi fleet [43]. The plot shows routes shared by two or more vehicles, that is somewhat popular routes. The right plot shows a complete graph of locations and routes connecting them, generated by the *maps2graphs* project [9] which generates all-to-all graphs from individual routes obtained from actual maps using Google Maps API. The specific graph shown in the figure is the one that collected around 400 locations of FamilyMart convenient stores in northern Kyushu. The choice of FamilyMarts is valid in view of the *density center*

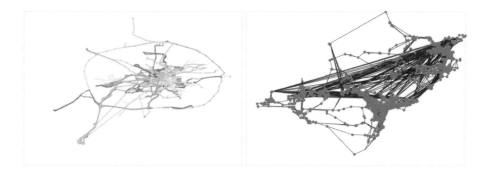

Figure 7. The Rome Taxi Crawdad versus the Kyushu FamilyMarts Maps2Graphs datasets.

research which states that distribution of people (and car) density roughly follows the distribution of main objects of accompanying infrastructure such as supermarkets, smaller stored, gas stations, etc. [3].

Both the above traces create problems when trying to apply them directly in models of vehicular clouds.

For the *Rome taxi* trace the problem is in the small size of the fleet. While it is still possible to define popular routes (as the above plot shows), it is completely impossible to find multiple taxies that park together. This means that it is impossible to extract density centers from this trace.

The *Kyushu FamilyMarts* trace intrinsically contains density information, which is not found in numeric form in the trace itself but can be easily calculated using the algorithms discussed further in this chapter. The notion of *popular routes* is missing from the trace—which is the main problem with it—but can be generated using density information and the raw routes in the trace. The generation process is shown further in text. Note that this trace has additional useful information such as individual legs of routes between nodes with travel time and distance provided by Google Maps API. This numeric information is used for analysis further in this chapter.

Having considered the two above options, it was decided to use the *Kyushu FamilyMarts* trace for analysis.

Figure 8 shows a larger set of modeling components, divided into real (trace-based) (right) and synthetic parts, with some links between the two. The rest of this subsection discusses each component in detail.

Mobility Traces describe real traces like the Rome Taxi [43] or Kyushu FamilyMarts *maps2graphs* traces above. Note that the nature of the two traces are different. Rome Taxi trace is the mobility trace where data unit is a GPS coordinate. Kyushu FamilyMarts trace is a kind of GIS trace but has additional elements which can describe mobility. Real traces are essential for recreating realistic conditions in simulation.

Density Centers is a real component, where the Kyushu FamilyMarts trace is a specific example [9]. Having a real GIS trace, it is not difficult to process it into a density map. Description of Voronoi plots below revisits this subject. Density distributions in analysis further in this chapter are used to distribute vehicles geographically. The logic linking density with distribution logic is straightforward—locations with higher

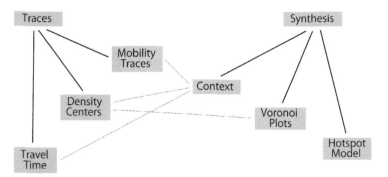

Figure 8. Modeling concepts related to simulations of vehicular clouds.

density have higher ratios of populations of humans. Since vehicles are owned by humans, vehicle distribution mirrors that of humans.

Travel Time is related to density centers—example of its use in practice can be seen in [3]. Travel time normally comes in form of a distribution of travel time against its duration. The relation to density centers is in the part where commutes in large cities (same for Fukuoka and most cities in Japan) are mostly between less populated and more populated areas. Travel time distribution helps pick the specific nodes in the graph which fit the distribution.

Voronoi Plots is a well-known method which can be used to convert maps of locations into the density map. The logic to Voronoi plots is simple—each node is allowed to increase its area until it clashes with the areas from other nodes. These clashes gradually reveal borders between nodes. When the algorithm completes, one can convert the areas into density by simply taking their relative values. Note that the relation here is inverse, i.e., larger areas correspond to lower density. This exact algorithm is applied to the Kyushu FamilyMarts trace to produce the density map further in this section.

Hotspot Model is fully described in [8]. It defines distributions where large outliers are rare but extremely large. The hotspot distribution can be applied to distribution of parking areas within each node on the map. In practice, such descriptions can effectively model the difference between tenant and one-story houses, large parking areas and individual cars randomly parked outside, etc.

Context is a vaguely defined overall framework for each specific model. It provides context for most other parts of the model and can answer the following example questions:

- what is described by density in the density map?
- how to place mobility traces in context—people commuting to and from work [3]?
- given a distribution of travel time, between which locations should one assign each individual travel [3]?

Analysis further in this chapter cannot physically accommodate the full model above and is forced to introduce some simplifications. Specifically, travel time and hotspots are not used in the analysis. However, daily commutes between out-city and in-city nodes, grouping of vehicles within each node are considered in full. Travel time distribution is replaced with a simplified element which assumes that a portion of vehicles that parks together overnight, commute together in the morning. Detailed description of the simplified model is offered further in this chapter.

4. Practical Scenarios and Analysis

This section uses a simplified version of the model discussed in the previous section for performance analysis. As was mentioned above the choice was made in favor of the Kyushu FamilyMarts trace which, although is not strictly a mobility trace, is found to be more suitable for analysis because it can be converted into a density map and thus support realistic distribution of vehicles across all the nodes in the map. The first subsection explains all the simplifications made to the general model described above. This section also defines the storage and sensor clouds in the context of the simplified simulation model and analyses their performance.

4.1 The Simplified Model

As was mentioned above, simplifications are necessary to reduce the volume of analysis to a reasonable size. When volume is not an issue, the below simplified model can be enhanced in the following way (these can be considered as a recommendation which can enhance the model at the expense of increased complexity):

- hotspot distribution can be applied to each node in the graph, thus describing distribution of the allocated sub-fleet of vehicles across tenant and one-story houses at the respective location; note that different hotspot models might be necessary depending on the distance of a given location from the city;

- travel time distribution can be applied to increase detail when selecting which vehicles travel from which locations to which other locations—in the simplified model each sub-fleet is simply divided into commuting versus non-commuting parts.

Figure 9 shows the simplified model that is best represented as a sequence of *forks* that divide the fleet into smaller sub-fleets, each time using only one of the two parts for simulation. The following are the three forks:

- if cloud is installed or not, as a simple way to set the ratio of the total number of vehicles in the city that implements the cloud platforms and can therefore be used to run cloud services;

- the cloud-aware sub-fleet is further divided into the part that can support grouping and which are too far from other cloud-aware vehicles to form groups; the selected group is assumed to be *parked together* (not all in one place but many smaller groups can be considered as one large group from the viewpoint of the cloud service) and is geographically distributed in accordance with the density map;

- not all the vehicles that park together commute to city for work, the alternative is the group that is assumed to conduct irregular or local travel (or, shortly, non-commuting part).

Figure 9 shows that division is continued after the *moving together* sub-fleet. The further divisions are natural and come from the fact that routes from a given parking space to different (even in-city) locations may not always overlap. In this case, the moving together sub-fleet can be further divided into groups that share some legs on the routes but not the entire routes. Further complicating this point, vehicles from other nodes on the graph (locations) may share routes, in which case the *travel together* sub-fleet experiences further changes depending on each location.

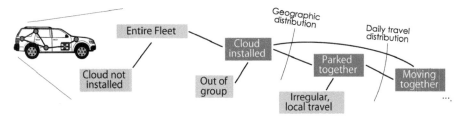

Figure 9. A simplified model of a fleet of vehicles and cloud services they support.

Continuing on further divisions, travel time has to be the same for all vehicles that are assumed to be traveling together and can therefore become nodes in a sensor cloud.

All the above enhancements are part of the general model described above while the model in Figure 9 deals with them in the following way. The forks are defined by simple ratios from each subsequent sub-fleet, three values in total. The total fleet of sedan-class cars in Fukuoka prefecture is around 800 k vehicles. The *cloud installed* ratio is applied to the total population, next ratio is applied to the *cloud installed* sub-fleet, and so on. Ratios are simple but sufficiently descriptive. Smaller values directly correspond to situations where a very small sub-fleet at any given location participates in a given action. For example, the lowest ratio for the *travel together* sub-fleet is 5% which means that only 5% of vehicles that park in group at a given location are likely to commute together to work.

The forks in Figure 9 also correspond to actions on the part of the cloud service provider which progress in the following sequence:

- when cloud in installed at a given vehicle, the cloud service becomes aware of it and starts monitoring its activities considering its future including into either sensor or storage cloud, or both;
- if vehicles are parked together, and the cloud service confirms the consistency after several days, it can decide to start offloading content to its storage space, at which time the cloud software installed in the vehicle will take an active part in grouping with other vehicles while parking and optimization of syncing traffic in accordance with the GroupConnect concepts;

- for parked-together and other vehicles, the cloud service also monitors the consistent routes which are used to assign likelihood of a given vehicle to be at a given road at a given point in time, the data created in the process are used by the cloud service to decide which are the most reliable vehicles to get sensor data in realtime; the alternative to long-term planning can be runtime selection based on the monitoring GPS locations at runtime.

Note that the above modeling restricts usecases to those which *require* GroupConnect and are not willing to settle for the SingleConnect fallback. If the latter is allowed, then all the above restrictions are lifted and performance of such fleets can be viewed from the angle of relative improvements in response to improving conditions. However, since the main objective of this chapter is to show how vehicular clouds can benefit specifically from GroupConnect, analysis further in this section focuses only on cases when grouping of multiple vehicles is guaranteed.

4.2 The Real GIS Dataset

This section redraws the Kyushu FamilyMarts dataset, takes a smaller sample from it and generates the density map from the sample. The sample is taken to reduce the number of nodes from around 400 in the total dataset to under 200 in the sample.

As was explained above, density maps can be created using Voronoi plots as an optimization tool. The optimization is simply finding the low-energy border between nodes and using the relative areas as inverse of density.

Figure 10 shows the whole dataset and the sample that includes only the 70 km area around Fukuoka city. Red dots near the center in both charts marks the very center of the city. Bullet size denotes the relative area size between the nodes obtained using the Voronoi plot optimization. The bullet size becomes density by taking its inverse value and normalizing the entire set.

The right side of Figure 10 is used for analysis further in this section. The actual map is not shown to retain visual clarity. However, the shape of the plot can be explained. Fukuoka is a compact city built next to the ocean (upper part of the plot). The center of the city is next to the shore while the city itself extends along the shoreline on both sides. The lower part of the plot is the part if the city that extends in-land to several smaller cities.

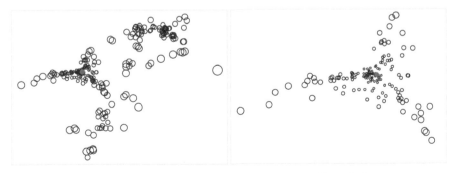

Figure 10. The subset of the maps2graphs dataset used for analysis.

For the simplified model we need a clear separation of in- versus out-city locations. In analysis they are used by the *travel-together* sub-fleet of vehicles to commute. Figure 11 is created in the following way. All the nodes are ordered by decreasing density. Top 25 nodes are then assigned as in-city nodes and bottom 100 nodes as out-city nodes. In the figure they are marked as circles versus rectangles. Additionally, to visualize the spread of density, out-city nodes are marked by gradually decreasing opacity in steps at 50, 75 and 100 nodes (the last group is the darkest black). The difference in opacity shows that all groups are spread geographically, meaning that density is not in linear relation to distance from city center.

Based on the results in Figure 11 one further simplification is possible. Since top 25 densest nodes are located in almost the same spot around the center of the city, efficiency can be improved by selecting only one node to represent all the in-city locations. The very center of the city is selected as such a node. Note that this simplifies analysis of shared routes because, while there are multiple departure points of commutes, there is now only one arrival point.

Note that this trick is applicable to Fukuoka but may not apply to other cities. For example, Tokyo has several centers scattered around a fairly large area which makes the above simplification impossible. In Japan, Fukuoka is known is one of the most compact cities.

Figure 11. In-city and out-city clusters of locations.

4.3 Analysis Setup

As was mentioned earlier, simulation is configured by three ratios that are applied to *cloud installed, park together* and *travel together* sub-fleets. *Cloud installed* ratio is selected from the set {1.0, 0.7, 0.5, 0.3, 0.1}, where 1.0 means that the entire fleet of 800 k vehicles have installed the cloud platform. By definition, only a fraction of vehicles can find successful grouping with others, which is why this ratio is selected from the set {0.5, 0.3, 0.1}. *Travel together* yet further diminishes the sub-fleet with the same set {0.5, 0.3, 0.1} of values, this time applied to the *park together* sub-fleet. Applying these ratios creates three sub-fleets with gradually decreasing sizes in each location. Note that geographical distribution is performed independently based on the density map discussed in previous sections.

One simulation run is executed as follows. First, a distinct configuration of ratios is selected and used to calculate fleet sizes for each location. Density map is used after the first ratio (cloud installed) to define the number of vehicles for each location. Note that, while distribution takes into account all the nodes in the graph, vehicles are assigned (observing the proportions) only to the nodes which are tagged as out-city nodes. As was shown earlier in this section, there are 100 out-city nodes, vehicles from which are assumed to commute to a single in-city node in the center of Fukuoka city. Having defined the number of vehicles assigned to each out-city location, the number is then further divided using the two other ratios.

At the initial state of the simulation, we already know the number of vehicles that can support the *storage cloud* service. It is sufficient to assume that all the vehicles at that location can be part of the storage service. *Sensor cloud* requires further effort.

The *travel together* sub-fleets from all locations are assumed to commute to the single in-city node in the morning and return back in the evening (work-day commute). It is interesting to visualize how many vehicles share which legs of which routes between pairs of nodes. This is not a straightforward calculation because shared legs are different depending on the specific combination of locations. To simplify this result, all routes are decomposed into individual legs and overlaps are estimated for each individual leg. This way, each leg gets its own number of vehicles, plus its length in meters and estimated travel time in seconds are read directly from the *maps2graphs* trace discussed earlier in this chapter.

4.4 Analysis Results

This section analyzes performance of the above simplified model in two plots. The first plot compares the best and the worst configurations in terms of what a given cloud service can expect when running on such a fleet. The second plot focuses on the traveling fleet (and therefore the sensor cloud service) and visualizes the effect that configuration parameters play on the ability of the cloud to collect sensor data while vehicles are in commute.

Figure 12 shows the best (upper) and worst (lower) situations. The best here is represented by the configuration {1.0, 0.5, 0.5} for the three ratios. The worst configuration is {0.1, 0.1, 0.1}, respectively. Each curve is a distribution that represents each of the two selected cloud services. Distributions represent vehicle count changing with increasing distance from the center of Fukuoka city. The curve for the storage cloud service starts from 5 km and continues until the edge of the graph (about 70 km). The curve for the sensor cloud has no bounds and is created naturally based on the actual distances of shared legs in routes from various locations.

In a way, the plots in Figure 12 have all the natural and, therefore, expected features. For example, shared legs on routes to in-city nodes can be very close to the

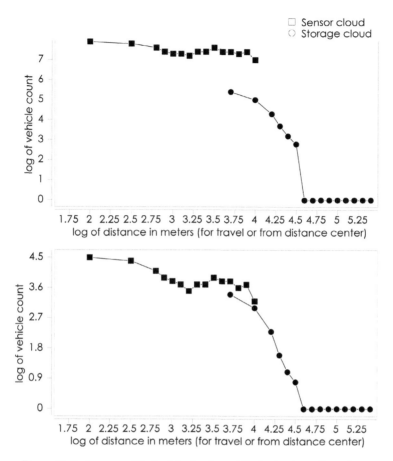

Figure 12. Performance at the best (top) and worst (bottom) ratio configurations.

center of the city but cannot extend forever because, as the plots show, there are no shared legs beyond 10 kms. This is an expected behavior.

The results in Figure 12 is as follows. The best case for storage cloud can offer fleets of between 10 k and 100 vehicles. The worst case can only offer between 700 and 10 vehicles for the entire fleet, depending on the distance. It can be argued that cloud provider would probably give up on trying to store content at the edge of such a fleet and would concentrate at locations much closer to the city. Using vehicular clouds for storage inside the city is arguably meaningless because cloud providers can rent cheap facilities inside the city.

For the sensor cloud the best case is at 100 k and the worst at 10 k vehicles, almost consistently for the entire effective distance from city center. This property comes directly from the trace which shows that within this range of distances, routes from many locations share the same legs, which means that cloud provider has a sufficient number of data sources to collect data from. Also, the high number of vehicles even in the worst case shows that even large GroupConnect sessions are not hard to achieve on shared legs of routes.

Figure 13 focuses only on the traveling cloud, i.e., the fleet that implements the sensor cloud service. Virtual axes on the plots focus on individual legs and display the length of the leg (in meters) and the travel time (estimated by Google Maps API). Horizontal scales show the sequence of configuration parameters, where each combination is always the *cloud/park/travel* sequence of ratios. Horizontal scale can be viewed as a gradually decreasing in size sub-fleets. To avoid crowding the plot, top 3 legs are shown in each plot.

The results in Figure 13 are as follows. The biggest shared legs are between 9 km and 13 kms, that corresponds to travel time between 600s and 900s. These numbers should be considered literally as time span that can be exploited by GroupConnect to receive (information support) or send (sensor cloud) large bulks of data to/from the cloud. It is interesting that almost for the entire range of configurations, this performance is unchanged, but there are intermittent minor drops when the *traveling ratio* is at 0.05. This is to be expected—when the traveling sub-fleet is too small, it means that many distant locations stop providing vehicles for morning and evening

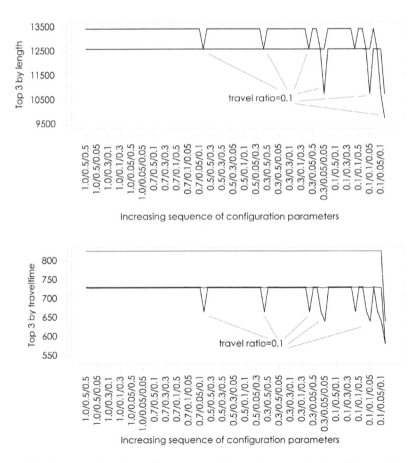

Figure 13. Performance of top 3 travel legs across all the combinations of ratios. Distance is in meters and time is in seconds.

commutes which causes some attractive (long) legs to disappear from shared routes. The same effect is repeated when all sub-fleets are small (cloud ratio at 0.1), in which case even top legs are very short both in distance and time.

The practical value of the data in Figure 13 is as follows. The road graph around Fukuoka city is such that there are many extended legs shared by routes from many locations. These legs are fairly stable and disappear only when the coverage area is reduced to the level when these legs disappear. The plot shows that this happens only in very restricted environments in which only a very small fraction of vehicles commute to the city for work. Taking specifically the case of Fukuoka, a much larger fraction of vehicles is part of regular morning and evening commutes.

Granted the above analysis is fairly simplistic and does not cover many other practical metrics (distance from city center, actual coverage area, etc.), it does show that both the selected cloud services are feasible for the entire area of the graph. The feasibility can be affected only at the most restricted configurations (very few commutes, etc.) which is arguably too pessimistic for Fukuoka specifically and most other large cities around the world, in general. Detailed analysis of all the related metrics will be offered in future publications on the subject.

5. Summary

This chapter is the first known attempt to introduced the existing concept of GroupConnect to vehicular clouds. The chapter makes a clear distinction between vehicular networks and vehicular clouds—the notion of connected vehicles playing the role of the border between the two. The notion of connected vehicles is the same in human systems where each smartphone has a continuous but low-throughput connection to the Internet. Just like in human systems, one of the useful applications of GroupConnect in vehicular clouds is its ability to boost throughput.

Throughput-boosting is not the sole goal of GroupConnect. The technology itself is defined as virtualization of resources of an arbitrary group of nodes (wireless, human or vehicular). Once virtualized, the resources can be optimized for many practical targets. When throughput-boosting is the target, the virtual node can pool all the Internet connections of individual members of the group into a single virtual Internet connection. A separate subsection in this chapter was dedicated to showing that such pooling of network connections is extremely easy to implement in context of cloud services, where access tokens from APIs can be shared between nodes and used independently. This basic technology can successfully boost throughput of a vehicular group to the level of, theoretically, the wireless technology used for peer-to-peer connections within the groups. If WiFi Direct is used—and it can be used in practice because it is available in many smartphones today—the ceiling for throughput is around 30 Mbps, which is very high compared to what has been shown in recent benchmarks on practical 3G/LTE capacity (400 kbps on average).

The goal of this chapter, however, was not to prove the utility of GroupConnect, repeatedly. The target of this chapter was to show that GroupConnect and the related self-organization techniques are the only option which can guarantee feasibility of vehicular clouds in practice. In this respect, this chapter mostly focuses on the specifics of applying GroupConnect in vehicular environments.

The chapter argued that uniqueness of vehicular GroupConnect starts from how the groups are formed. First, the difference in mobility patterns between cars and humans is a known fact. However, this chapter stresses that *parking* should not be left without consideration. As far as this chapter is concerned, the time each vehicle spends in a parking lot, next to other such vehicles, is the wasted resource which can be of high value to cloud service. Specifically, this chapter shows that parking lots can be turned into large virtual hard disks maintained remotely by cloud services. Similarly, vehicles which travel together at proximities sufficient for successful GroupConnect, can be treated by a cloud service as a huge sensor network.

Such technologies are not possible without proper fleet management. This chapter links the notion of fleet with the notion of population management which is already known in clouds. In fact, each vehicle in a managed fleet becomes an item is a new, vehicle-based, cloud population. In this respect, the two concepts are merged completely in this chapter.

Let us review the process of building the two cloud services—storage cloud and sensor cloud—from scratch, following all the discussions in this chapter.

First, the fleet has to exist in the first place. Natural aggregates like certain models of cars or all cars from a given maker, are ready-made fleets, by definition. Fleets can also be connected by a service they share—this chapter brings the example of the EV battery replacement service that exists in recent literature. In one way or another, fleets are not uncommon today, and some effort can be made to create bigger and cross-maker/cross-provider fleets in the future.

When the fleet is ready, each vehicle in the fleet has to install the cloud platform without which vehicular clouds are impossible. This chapter shows that the easiest way to accomplish this is to give your car a 3G connection. Existing attempts, notably the connected cars effort, are not sufficient because it would greatly reduce the time span when your car is connected to the Internet. While it is possible to control fleets in which vehicles appear online intermittently, cloud services would greatly prefer the always-on option. In fact, intermittent connectivity would make the above storage cloud usecase nearly impossible simply because cars would rarely be on and still (not in motion) at the same time.

With these two stages passed, the cloud now has the complete freedom of running cloud services in your vehicle. The labor is divided between the software that runs in your vehicle's cloud platform and the one that runs in the cloud itself. Your local cloud platform is in charge of GroupConnect and self-organization. The cloud itself plays only a small part in self-organization and mostly controls the sync between itself and the content stores or generated by the vehicle.

One major problem formulated by this chapter is the modeling component of vehicular clouds. Having discussed several existing examples and individual technologies, it was shown that vehicular clouds require a drastically different approach to be able to correctly analyze feasibility of cloud services running on large vehicular fleets. While this chapter performs feasibility analysis based on a simplified model, further work and academic discussion are needed in this area. Future publications on this topic will focus on improving the model and creating simplified versions which can help analyze specific usecases without loss of modeling accuracy.

References

[1] M. Zhanikeev. Opportunistic MultiConnect with P2P WiFi and Cellular Providers. Advances in Mobile Computing and Communications: 4G and Beyond, CRC (in print), 2015.

[2] M. Zhanikeev. Multi-Source Stream Aggregation in the Cloud. Chapter 10 in the Book on Advanced Content Delivery and Streaming in the Cloud, Wiley, 2013.

[3] M. Zhanikeev. On how smart cities can improve social utility of their citizens' commutes. IPSJ Journal of Information Processing, 22(2): 253–262, 2014.

[4] M. Zhanikeev. Wireless user: a practical design for parallel MultiConnect using WiFi direct in group communication. 10th International Conference on Mobile and Ubiquitous Systems: Computing, Networking and Services (MobiQuitous), 2013.

[5] M. Zhanikeev. Group connect in a new wireless University Campus. IEICE Technical Report on Smart Radio (SR), 114(165): 27–30, July 2014.

[6] M. Zhanikeev. Experiments with application throughput in a browser with full HTML5 support. IEICE Communications Express, 2(5): 167–172, May 2013.

[7] M. Zhanikeev. WiFi direct with delay-optimized DTN is the base recipe for applications in location-shared wireless networking virtualization. IEICE Technical Report on Radio Communication Systems (RCS), 113(386): 25–28, January 2014.

[8] M. Zhanikeev and Y. Tanaka. Popularity-based modeling of flash events in synthetic packet traces. IEICE Technical Report on Communication Quality, 112(288): 1–6, November 2012.

[9] M. Zhanikeev. Maps2Graphs: A socially scalable method for generating high-quality GIS datasets based on Google Maps API. IEICE Technical Report on Intelligent Transport Systems Technology (ITS), 113(337): 73–76, November 2013.

[10] M. Whaiduzzamana, M. Sookhaka, A. Gania and R. Buyya. A survey on vehicular cloud computing. Journal of Network and Computer Applications, 40: 325–344, 2014.

[11] Z. Alazawi, M. Abdljabar, S. Altowaijri, A. Vegni and R. Mehmood. ICDMS: An intelligent cloud based disaster management system for vehicular networks. Communication Technologies for Vehicles, Springer LNCS, 7266: 40–56, 2012.

[12] R. Yu, Y. Zhang, S. Gjessing, W. Xia and K. Yang. Toward cloud-based vehicular networks with efficient resource management. IEEE Networks, 27(5): 48–55, 2013.

[13] M. Gerla, E. Lee, G. Pau and U. Lee. Internet of vehicles: from intelligent grid to autonomous cars and vehicular clouds. Wireless Sensor Networks Magazine (in print), 2014.

[14] Y. Xu and J. Yan. A cloud based information integration platform for smart cars. 2nd International Conference on Security-Enriched Urban Computing and Smart Grid, pp. 241–250, 2011.

[15] A. Saha and D. Johnson. Modeling mobility for vehicular ad-hoc networks. 1st ACM international workshop on Vehicular ad hoc networks (VANET), pp. 91–92, September 2004.

[16] Connected Vehicle Cloud: Under the Hood, Ericsson Whitepaper, 2013.

[17] Cards Connect with apps, the Cloud at CES. Wired, 2012.

[18] IBM to Build Cloud-based Connected Cars Platform in China. Infortechlead, 2014.

[19] Jasper Connected Car Cloud Drives Connected Vehicle Innovation. Jasper Whitepaper, 2014.

[20] Nissan Taps Airbiquity for Cloud-Connected Cars. Telecoms, 2014.

[21] M. Satyanarayanan. Fundamental challenges in mobile computing. The Fifteenth Annual ACM Symposium on Principles of Distributed Computing (PODC), 5: 1–6, May 1996.

[22] N. Fernando, S. Loke and W. Rahayu. Mobile cloud computing: A survey. Elsevier Journal on Future Generation Computer Systems, 29: 84–106, 2013.

[23] M. Satyanarayanan, P. Bahl, R. Caceres and N. Davies. The case for VM-based cloud lets in mobile computing. IEEE Pervasive Computing 8: 14–23, October 2009.

[24] L. Deboosere, P. Simoens, J. Wachter, B. Vankeirsbilck, F. Turck, B. Dhoedt and P. Demeester. Grid design for mobile thin client computing. Future Generation Computer Systems, 27: 681–693, June 2011.

[25] O. Amft and P. Lukowicz. From backpacks to smartphones: past, present, and future of wearable computers. IEEE Pervasive Computing 8: 8–13, July 2009.

[26] G. Huerta-Canepa and D. Lee. A virtual cloud computing provider for mobile devices. 1st ACM Workshop on Mobile Cloud Computing and Services: Social Networks and Beyond (MCS), 6: 1–5, 2010.

[27] R. Balan, M. Satyanarayanan, S. Park and T. Okoshi. Tactics-based remote execution for mobile computing. 1st ACM International Conference on Mobile Systems, Applications and Services, pp. 273–286, May 2003.

[28] J. Flinn, S. Park and M. Satyanarayanan. Balancing performance, energy, and quality in pervasive computing. 22nd IEEE International Conference on Distributed Computing Systems, pp. 217–226, July 2002.

[29] E. Cuervo, A. Balasubramanian, D. Cho, A. Wolman, S. Saroiu, R. Chandra and P. Bahl. Maui: making smartphones last longer with code offload. 8th ACM International Conference on Mobile Systems, Applications, and Services (MobiSys), pp. 49–62, June 2010.

[30] E. Marinelli. Hyrax: cloud computing on mobile devices using MapReduce. Masters Thesis, Carnegie Mellon University, 2009.

[31] D. Huang, X. Zhang, M. Kang and J. Luo. Mobicloud: building secure cloud framework for mobile computing and communication. 5th IEEE International Symposium on Service Oriented System Engineering (SOSE), pp. 27–34, June 2010.

[32] M. Rumney. LTE and the Evolution to 4G Wireless, Design and Measurement Challenges. Wiley, 2013.

[33] X. Zhang and X. Zhou. LTE-Advanced Air Interface Technology. CRC Press, 2013.

[34] E. Hossain, M. Rasti, H. Tabassum and A. Abdelnasser. Evolution toward 5G multi-tier cellular wireless networks: An interference management perspective. IEEE Wireless Communications, 21(3): 118–127, June 2014.

[35] S. Glisic. Advanced wireless networks: Cognitive, Cooperative and Opportunistic 4G Technology. Wiley, 2009.

[36] A. Konak, G. Buchert and J. Juro. A flocking-based approach to maintain connectivity in mobile wireless ad hoc networks. Journal of Applied Soft Computing, 13(2): 1284–1291, February 2013.

[37] W. Ting and Y. Chang. Improved group-based cooperative caching scheme for mobile ad hoc networks. Journal of Parallel and Distributed Computing archive, 73(5): 595–607, May 2013.

[38] S. Zachariadis, C. Mascolo and W. Emmerich. Satin: a component model for mobile self organisation. On the Move to Meaningful Internet Systems (CoopIS/ODBASE), Springer LNCS, 3291: 1303–1321, October 2004.

[39] N. Banerjee, S. Agarwal, P. Bahl, R. Chandra, A. Wolman and M. Corner. Virtual compass: relative positioning to sense mobile social interactions. 8th International Conference on Pervasive Computing, pp. 1–21, May 2010.

[40] A. Vasilakos, Y. Zhang and T. Spyropoulos. Delay Tolerant Networks: Protocols and Applications. CRC Press, 2011.

[41] A. Balasubramanian, B. Levine and A. Venkataramani. DTN Routing as a Resource Allocation Problem. SIGCOMM, pp. 373–384, October 2007.

[42] CRAWDAD Repository of Mobility Traces. [Online]. Available at: http://crawdad.cs.dartmouth.edu (retrieved July 2014).

[43] R. Amici, M. Bonola, L. Bracciale, P. Loreti, A. Rabuffi and G. Bianchi. Performance assessment of an epidemic protocol in VANET using real traces. International Conference on Selected Topics in Mobile and Wireless Networking (MoWNeT), 2014.

10

MAVSIM: Testing VANET Applications Based on Mobile Agents

*Oscar Urra** and *Sergio Ilarri*

ABSTRACT

Currently, intelligent vehicles and Intelligent Transportation Systems (ITS) are attracting a great deal of interest in research. Moreover, it is probable that in the next few years this interest will increase, because of its encouragement from instituitions of America and the European Union (through regulations) and the widespread adoption of vehicular communication systems, as well as the possibilities offered by this kind of networks: applications related to driving safety, entertainment, collection of data from the environment, social applications, etc. Ad-hoc vehicular networks (VANETs) have a special interest, since they operate using P2P (peer-to-peer) communications among their users, with no need for a centralized infrastructure, and thus can be deployed more easily and with less cost. Users that want to participate in the VANET would only have to install a small device in their cars and they will immediately join the network.

However, there are a number of issues that can slow down the adoption of such kind of networks. First, the potential high number of users will increase the number of wireless communications with the risk of saturating the available bandwidth, which has motivated the development of the so-called cognitive radio networks. Another difficulty, that is inherent to ad-hoc networks, is the transportation or routing of the data from one node (vehicle) to another in such a changing environment, where all the nodes are constantly moving and the connections among them last at most a few seconds. Finally, another key difficulty is how to design and develop applications that can benefit from the

University of Zaragoza (Zaragoza, Spain).
*Corresponding author: ourra@itainnova.es

advantages provided by this kind of networks, since both their development and testing can be complex. Mobile agents are programs that have the capability to move their execution from one computer or device to another, and they are considered an interesting technology for VANETs. In this chapter, we present a simulation approach that we have developed, focused on the evaluation of data management strategies for vehicular networks that are based on the use of mobile agent technology.

1. Introduction

Vehicular ad hoc networks (*VANETs*) [20], which are a highly-dynamic type of *mobile peer-to-peer networks* (*mobile P2P networks*) [2], are expected to become a reality in the coming years and numerous useful applications for drivers and passengers will be developed. However, VANETs are also likely to present significant challenges from the point of view of data management [5, 6, 7], related to the acquisition, processing and the efficient and effective exchange of data among vehicles. Beyond the traditional limitations of mobile computing scenarios, many of the difficulties are a consequence of the use of short-range wireless communication technologies upto about 200 meters, which are unstable, subject to frequent disconnections (network partitioning), have limited communication range, and are subject to security attacks (since the transmitted data are broadcasted). Moreover, the increasing number of wireless devices, and its high concentration in the same geographic area, reduces the available bandwidth since all that devices share the same limited radio frequencies for the transmission of data. Thus, cognitive radio networks [1, 9, 10, 15, 19, 30] could play an important role.

From our point of view, the use of mobile agent technology [13, 18, 24] in this environment is a research avenue worth exploring. Mobile agents are autonomous software entities that have the ability to transfer themselves from one execution environment to another by using a communication network. We believe that mobile agents could play an important role as a middleware for the development of applications for vehicular networks, as they naturally support disconnected operations and distributed data management. Overall, they can be an interesting design abstraction in a variety of scenarios.

If we focus on the test phase of the development of applications and data management strategies for vehicular networks, it is clearly very expensive and inconvenient to perform software testing using real devices in real cars. So, real-world tests are usually limited to very small and controlled scenarios, mainly as a proof of concept or to obtain measures that can be used to fine-tune some simulator. The most logical alternative is indeed to simulate the behavior of the developed software in a controlled and simplified environment that can be managed more easily and that supports a large-scale simulation with a high number of vehicles. There are a variety of simulation software to test and analyze with the required precision different aspects of communication networks and vehicular traffic, e.g., [16, 17]. However, as far as we know, none of the most popular simulators can be easily used to simulate scenarios with the goal of evaluating data management solutions based on the use of mobile agents. Motivated by this, we have developed a vehicular network simulator in which mobile agents can also be simulated, as well as a set of tools intended to facilitate the analysis of the results obtained in a variety of scenarios.

This chapter extends the work presented in [28], including among others an experimental evaluation, more detailed and improved descriptions, and some considerations concerning cognitive networks. The rest of this paper is structured as follows. In Section 2, we briefly describe the technological context of our research area. In Section 3, we focus on the development of our own simulator, describing its architecture and functionalities. In Section 4 we evaluate the simulator performance by testing some parameters. Finally, in Section 5, we summarize our conclusions and outline some lines of future work.

2. Technological Context

In this section, we have provided an overview of the technological context that serves as a background for this work, highlighting those aspects that are particularly relevant for the simulation approach that we will describe along the chapter. First, we overview basic aspects of VANETs and vehicular cognitive networks. Afterwards, we focus on mobile agent technology and the potential interest of using such a technology in vehicular networks. Finally, we discuss some popular simulators for VANETs.

2.1 Cognitive Vehicular Networks

Using VANETs in vehicular applications has a number of advantages over a traditional client-server approach, such as: (1) dedicated centralized support infrastructure (expensive to deploy and maintain) is not required; (2) users do not need to pay for the use of these networks; and (3) it allows a very quick and direct (i.e., without intermediate proxies or routers) exchange of information between two vehicles that are within range of each other, which may be crucial for safety applications for vehicular networks. Moreover, many application scenarios do not need to communicate with a specific target vehicle but with all the vehicles within a certain area, and therefore broadcast (the typical communication mechanism in VANETs) is a suitable choice.

Other communication schemes can also be considered, based on a fixed infrastructure or mobile telephony networks, e.g., 3G/4G. Thus, even if it is unrealistic to assume the availability of a generalized wide-area fixed infrastructure in the near future, mobile telephony networks already offer new perspectives for the development of applications to assist drivers. Such solutions, based on a centralization of the data and decision processes, still suffer from issues such as poor scalability or low reaction time available when dealing with some events like an emergency braking. So, although it is important not to rely on a fixed network infrastructure, which can be difficult and expensive to deploy at a large scale and with global availability, some roads could also offer static relaying devices which provide internet access to nearby vehicles by using a fixed network (enabling *vehicle-to-infrastructure-V2I-communications*).

VANETs open up a wide range of opportunities to develop interesting systems for drivers. Although safety applications are usually emphasized, there are also interesting applications related to comfort, entertainment, and travel efficiency. However, a number of difficulties can also arise. Some of these difficulties are due to the fact that two vehicles can communicate directly only if they are near each other (the range of the wireless communication devices may be limited in practice to about 100–200 meters), which leaves a short period for potential data exchange. Moreover, multi-hop routing protocols may be needed to reach a distant vehicle. The development of applications for vehicular networks requires taking these constraints into account. They may also need to consider other special features of this dynamic context: the geographic area of interest could spread over a large extension, users (usually, the drivers) can move constantly at different speeds and directions (sometimes according to predefined routes, but not always), the density of the network nodes can vary depending on the place or the time of the day, etc.

An additional difficulty for VANETs, and for any other type of network using wireless communications, is the potential saturation of the communication spectrum due to a high concentration of devices within the same area. Wireless communications share the same medium (a range of radio frequencies grouped in *channels*), which cannot be used at the same time by all users, as there would be interfere in the communication. Therefore, it is necessary to have a medium access control protocol [3] that allows a communication device to obtain the designated range of frequencies allocated for exclusive use during a limited amount of time, while the rest of devices wait for their turn to transmit. When there are a high number of transmission-capable devices in the area, the available transmission windows are smaller, causing slower data transfers and higher latencies.

The ranges of radio frequencies available for data transmission are usually divided into separate sub-ranges for *licensed* and *unlicensed* users. Unlicensed communication devices are not allowed to use the licensed frequency ranges, even when there is no licensed device and the corresponding frequencies are unused. Due to this inefficiency, the concept of *cognitive radio* (*CR*) [19] was born, enabling better use of the available spectrum. Thanks to recent regulations made by telecommunication authorities such as the FCC (Federal Communications Commission) concerning the use of the radio-electric space [12], in this type of radio transmissions unlicensed devices can use licensed ranges as long as they are not being used at that moment by any licensed device [22]. In case a licensed device starts transmitting, it will take precedence over unlicensed devices. Unlicensed devices would switch immediately to unlicensed frequencies in order to leave the licensed spectrum free for its use by the higher-priority (licensed) clients.

To accomplish this, there are several methods [30], such as sensing the possible presence of licensed devices before trying to transmit or, if the communication device does not have this ability, the use of constantly-updated databases (provided by the telecommunication authorities) that contain temporal and spatial information about the licensed frequencies that are expected to be used [1]. Additionally, the different devices in the network (both licensed and unlicensed) can cooperate by exchanging information among them to optimize the use of the different available channels to increase their usage efficiency [10, 15].

The use of cognitive radio with these and other methods for managing the spectrum usage alleviates some of the problems described about network transmissions in vehicular networks, although it requires a more careful setup of the system and its components.

2.2 Mobile Agents

Mobile agents are software entities that run on an execution environment (traditionally called *place*), and can travel autonomously from *place* to *place* (within the same computer or between different computers) and resume their execution at the target [18]. Thus, they are not bound to the computer/device where they are initially created and they can move freely among computers/devices. To be able to use mobile agents it is necessary to execute a middleware known as a *mobile agent platform* [24], which provides an environment where they can execute as and can move to other execution environments and other services, e.g., communication, security, persistence, etc. This platform must be executed on every computer or device that could host agents in the distributed system, and it must offer some interface to communicate with other platforms present in other devices or computers. Key functions provided by a mobile agent platform is its ability to locate other available platforms and transmit the code of an agent from the origin to the destination of the agent's movement.

Thanks to the mobility capability of mobile agents, it is easy to build complex distributed applications that are at the same time flexible. Thus, a mobile agent can carry a required task wherever it is needed. If the task to be executed by an agent needs to be changed in the future, a new version of the agent (a new agent implementation) can be delivered. Thus, there is no need to keep specialized software installed on the computers/devices composing the distributed system: only the generic mobile agent platform software is needed and an agent implementing the required behavior can move there at any time.

Mobile agents can be designed and programed to provide useful benefits, e.g., autonomy, flexibility, and effective usage of the network, that make them very attractive for distributed computing. Motivated by the increasing popularity of mobile devices, mobile agents have been found useful for the development of applications in mobile environments. A mobile environment has a number of special properties, such as the need to rely on wireless communications due to the mobility of the mobile devices, that create a scenario completely different from that of a traditional distributed environment with fixed networks. Such an environment has a number of advantages, e.g., the processing is not tied to a fixed location, but also some drawbacks such as the limited computational power of mobile devices and the communication constraints imposed by the use of wireless communications (that usually either offer a low bandwidth, a high latency, and intermittent/unreliable connectivity, or are expensive or not available everywhere).

The autonomy, intelligence, and movement capabilities of mobile agents render them a powerful and flexible tool to build distributed systems, especially in mobile environments. For example, a mobile agent could be programed to visit certain devices, e.g., mobile devices embedded in vehicles in a dynamic network, e.g., a vehicular ad hoc network whose nodes are mobile devices and once it reaches devices storing relevant data to process the local data available there in order to collect new interesting data, and finally to return to the origin with the final relevant data collected. So, mobile agents can move the processing to the data source instead of bringing all the data to the node that will perform the processing, thus reducing the amount of data communicated through data filtering on-site. Mobile agents can also support disconnected operations, as an agent can live outside its *home device*, which can be turned off while the agent is performing its tasks elsewhere. Another interesting advantage is their capability to exploit the most suitable resources available, e.g., using powerful fixed computers instead of the limited resources of a mobile device, when appropriate. Finally, mobile agents can minimize network connections; thus, as opposed to a traditional client/server approach, that requires a connection open and alive while the request is being performed, an equivalent request processing using mobile agents would only require the connection active during the movements of the agents.

2.3 Using Mobile Agents in VANETs

As we have seen, VANETs offer useful opportunities for the development of applications for vehicles. However, they also cause some difficulties. For example, in VANET applications, data may need to be transported from vehicle to vehicle in order to reach locations that are not directly accessible due to the short range of the wireless communications used. Two major problems arise. First, as the propagation of the data can be slow, the information can be outdated when it reaches its destination. Second, it can be difficult to determine the destination itself and how to reach it. Given the highly mobile nature of a VANET, the destination could be a single specific vehicle, every vehicle present in a geographic area, all those vehicles matching a certain condition, etc.

To deal with these drawbacks, mobile agents can be very useful because of their adaptability and mobility features. Thus, *they can bring a processing task wherever it is needed*, and *the algorithm or agent's logic can be changed at any time* by deploying new versions of the agent's code. This flexibility is quite interesting in a vehicular network. A mobile agent-based application for a vehicular network can be updated by just releasing new versions of the involved agents, without the need to upgrade the software system of all the vehicles.

Another important advantage of mobile agents in vehicular networks is that *they can move to wherever the data are located in order to process and collect only the relevant data* (filtering out data which may be unnecessary). For example, if we want to obtain information from vehicles located within a certain geographic area, a mobile agent could move there and process the data locally. Once the most relevant data are obtained, they will be carried along with the mobile agent, keeping their size smaller (irrelevant data are discarded), and making it easier to transmit them in a scenario where communications could be constrained.

Finally, we argue that mobile agents can be *very useful for data dissemination* in vehicular networks. Thus, they can adapt easily to changing environmental conditions in order to improve the dissemination. For example, a basic flooding dissemination protocol will fail if the traffic density of vehicles is low and there are not enough vehicles to re-diffuse the data towards the target, as well as other problems such as the *storm broadcast* (network overloading). Other dissemination protocols, such as carry-and-forward [29], where the vehicles may hold the data to be transmitted until these data can be relayed to other vehicles, can be used in the case of low-traffic density.

However, considering the variety of existing dissemination protocols (and others that could be developed in the future), mobile agents seem an ideal technology to implement flexible and dynamic dissemination approaches and take suitable routing decisions. In this way, an agent can carry data and decide where and when to move, whether it should wait in the current vehicle before jumping to another one, whether it could be beneficial to clone itself, etc. With this approach, the routing decisions lie with the data themselves (encapsulated in the mobile agents) and different dissemination protocols (dynamic and adaptive to the current conditions) can be implemented.

2.4 Simulators for VANETs

A significant number of network and traffic simulators have been developed, both commercial and free or open source. In this section, we briefly present some of the most popular ones. An interesting recent survey on simulators for VANETs can be found in [16, 17]. Mobility models are studied in [11, 14, 23].

Network Simulators

Network simulators allow the configuration and simulation of detailed parameters of the devices and the communication process. For example, the technology used for data transmission (wireless, copper wire, fiber optic, etc.), the data loss ratio, latencies, shadowing effects[1] that make wireless communications more difficult, distance attenuation, etc. Some of the most used simulators of this type are *NS-3*[2] and *Qualnet* (*Quality Network*).[3] In our simulation approach we simulate communication networks from a high-level perspective, so we will not describe these simulators in more detail.

Traffic Simulators

Traffic simulators are specialized programs tracing the movement of vehicles, following different patterns and behaviors in different scenarios. Some examples of such simulators are SUMO and VanetMobiSim.

SUMO (*Simulation of Urban MObility*)[4] is an open source microscopic road traffic simulator. It allows the simulation of vehicles as single entities, with the ability of traveling through specific routes, changing the road lane, and following the traffic rules. It can handle scenarios with large road networks and a high number of vehicles. It can be enhanced with plugins and can interoperate with other software both by importing and exporting data using different file formats.

VanetMobiSim[5] focuses on vehicular mobility, and features realistic automotive motion models at both macroscopic and microscopic levels. At a macroscopic level, it can import maps from the US Census Bureau database, or randomly generate them using a

---

[1] Attenuation of the communication signal due to obstacles or interferences.

[2] https://www.nsnam.org/

[3] http://web.scalable-networks.com/content/qualnet

[4] http://sumo-sim.org/

[5] http://vanet.eurecom.fr/

Voronoi tessellation. It has also support for multi-lane roads, separate directional flows, differentiated speed constraints, and traffic signs at intersections. At a microscopic level, it implements different mobility models, providing realistic car-to-car and car-to-infrastructure interaction. According to these models, vehicles regulate their speed depending on nearby cars, they can overtake each other, and they act according to traffic signs in the presence of intersections.

Hybrid Simulators

A hybrid traffic-network simulator can simulate both traffic and network elements in a geographic scenario. Some examples of such simulators are NCTUns, EstiNet, and VEINS.

NCTUns (National Chiao Tung University Network Simulator)[6] is an extensible network simulator and emulator capable of simulating various protocols used in both wired and wireless IP networks, as well as wireless vehicular networks (including V2V and V2I communications), multi-interface mobile nodes for heterogeneous wireless networks, IEEE 802.16(e) mobile WiMAX networks, IEEE 802.11(p)/1609 WAVE wireless vehicular networks, various realistic wireless channel models, IEEE 802.16(j) transparent mode and non-transparent mode WiMAX networks, etc. It has been used for modeling VANETs and other ad-hoc networks as well as for the evaluation of real-life P2P applications and traffic signal control algorithms.

EstiNet[7] is a commercial product considered as the commercial version of NCTUns. It consists of an extensible network simulator and emulator capable of simulating various protocols used in both wired and wireless IP networks, as well as wireless vehicular networks (including V2V and V2I communications), among others. Regarding its traffic simulation capabilities, it can simulate multi-lane road networks, it incorporates different microscopic vehicle mobility models, and the behavior of any vehicle can be changed as it receives messages from the vehicular network. It has been used for modeling VANETs and other ad-hoc networks as well as for the evaluation of real-life P2P applications and traffic signal control algorithms.

VEINS (Vehicles in Network Simulator)[8] is an open source software that supports online re-configuration and re-routing of vehicles in reaction to network packets, it supports different vehicular mobility models, and relies on detailed models of IEEE 802.11p and IEEE 1609.4 DSRC/WAVE network layers (including multi-channel operation, QoS channel access, noise and interference effects). It can import scenarios from OpenStreetMap,[9] including buildings, speed limits, lane counts, traffic lights, and access and turning restrictions. It can also employ validated and computationally inexpensive models of shadowing effects caused by buildings as well as by vehicles. Finally, it supplies data sources for a wide range of metrics, including travel time and vehicle emissions.

---

[6] http://nsl.csie.nctu.edu.tw/nctuns.html
[7] http://www.estinet.com
[8] http://veins.car2x.org
[9] http://www.openstreetmap.org

It is also interesting to mention that the use of videogames to facilitate the evaluation of data management approaches for vehicular networks has also been proposed recently (see the *VANET-X* videogame).[10]

3. Testing with the Simulator

Network simulators have very limited (or non-existing) capabilities to simulate moving objects such as vehicles, whereas traffic simulators cannot simulate network communications, although some of them can export mobility traces to be imported later in a network simulator. A hybrid simulator can simulate both aspects at the same time.

However, none of the simulators mentioned can directly support the simulation of mobile agents, and so they cannot be easily used to evaluate data management approaches based on mobile agent technology. As this is the focus of our research, we were compelled to develop our own simulator (*MAVSIM, Mobile Agents in VANETs SIMulator*), that offers interesting functionalities in that context. The simulator has been developed in a generic way, with several configurable parameters and a modular and extensible architecture, to facilitate its use in a variety of scenarios to test different data management approaches.

In the rest of this section, we describe the main features of the simulator developed, its architecture, and a use case. The web site of the simulator (`http://sid.cps.unizar.es/MAVSIM/`) offers additional information, screenshots, videos, and can be downloaded.

3.1 Features

The key contribution of MAVSIM over other existing simulators is that it has been developed to enable an easy integration and testing of applications and data management protocols based on the use of mobile agent technology. The main features of MAVSIM are the following:

- It is written in Java, which makes it portable among different architectures and operating systems. It has been successfully tested in Microsoft Windows 7 and 8 (32 and 64 bits), Linux (32 and 64 bits), and Solaris (64 bits).

- It can run both in graphic interactive mode with a Graphical User Interface (GUI), or in batch mode (useful to execute a large number of simulations or experiments easily).

- Simulations can be recorded and replayed later (with step-by-step, pause, and rewind-and-forward functionalities), which facilitates a careful analysis of the whole process. For this purpose, the replay tool shown in Figure 1 is used; in that screenshot, we also show in the screen some information that is relevant for a sample monitoring task scenario.

---

[10] http://sid.cps.unizar.es/Vanet-X/

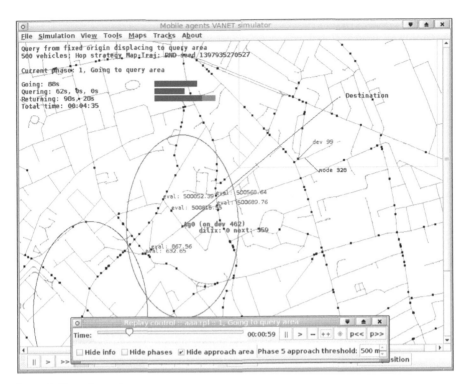

Figure 1. The replay tool.

- Any road map can be downloaded from OpenStreetMap. There is no limitation in the type of layout imported (cities, highways, rural areas, etc.). Figure 2 shows a scenario of an urban area and Figure 3 shows a scenario of a rural area. In the rural area, the density of roads is relatively low and they are surrounded by fields and forests.

- The simulated mobile agents can be programmed in a similar way as they would be programed using real platforms such as SPRINGS [13]. Thus, we provide methods such as *moveTo(targetDevice)* to simulate the movement actions performed by the agents. A built-in generic mobile agent platform is included in the simulator with the most common methods and primitives necessary for this.

- It can import traces generated by real vehicles or by other traffic simulators, and use them for the experiments with mobile agents.

- Road side units (RSUs), or fixed communication devices along the roads, can be simulated. So, not all the communication devices are necessarily linked to moving vehicles; instead, it is also possible to simulate other devices with a fixed location and a predefined communication range. With a configuration option it is possible to indicate whether these fixed devices are connected among them using a fixed high-speed network or not.

- It includes a variety of movement algorithms for the simulated vehicles, such as *Random way-point, Gauss-Markov*, and others, as well as movements following real GPS traces.

Figure 2. Example of scenario map for an urban area.

- In urban areas it can simulate the presence of buildings that block the wireless signal impeding communication. For this purpose, the geometric method described in [16] is used.
- The conditions of initial scenario for experiments can be set randomly or by using a known *seed*, which allows reproduction of similar conditions and repeat exactly the same experiment (with the same trajectories for the vehicles and other random conditions) if necessary.
- The simulator updates the status in fixed periods called *iterations* (by default, each iteration represents 1 second in the real world). Thus, the simulator is not a real-time simulator; instead, it updates the simulated state of the world in discrete steps. This has the advantage that, there is high simulation overload, the world simulated is consistent with the desired status of the real world (in other words, it is not possible to miss events, as there is no need to execute actions in a specific amount of time, i.e., in real-time).
- The simulator focuses on high-level aspects (e.g., communication delays) rather than on the simulation of low-level communication network features (e.g., the details of Medium Access Control protocols). For example, it is possible to set the latency for an agent's trip or the rate of communication errors. By appropriately setting high-level features we could emulate the effect of different conditions of cognitive networks [9].

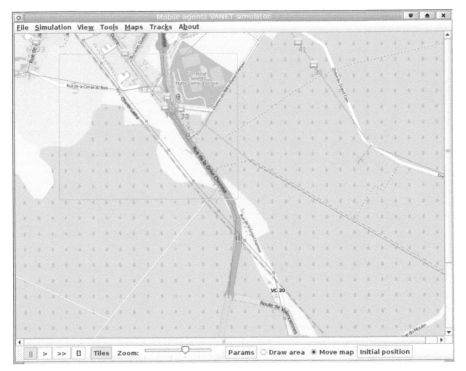

Figure 3. Example of scenario map for a rural area.

A number of parameters can be set to configure in detail the simulation scenario. Table 1 shows a summary of some of the general parameters that can be configured, as well as parameters that are applicable for the specific case of the evaluation of monitoring approaches based on mobile agents (see Section 3.3). Most parameters, when omitted, take a default value; others are optional and do not have a default value.

3.2 Structure and Classes

The simulator is designed to be extensible. It contains a number of classes with different functionalities (see Figure 4), and the classes are organized in modules (*packages*): one package contains everything related to graphics, another one the functionalities related to mobility strategies, etc. In this way it is easier to maintain the code and perform changes. Moreover, to facilitate its extensibility, some abstract classes and interfaces have been defined, that permit the adding of new functionalities in a quick and easy way. For example, if it is necessary to add a new mobility strategy for vehicles, there is an abstract class *Mobility* that includes the necessary methods and attributes for all the mobility strategies and makes adding a new one easier: it would be enough to define a new subclass implementing the methods of *Mobility* and extend the configuration files to add support to select and use that new mobility strategy in the simulations. In the same way, there is an abstract class to read different file formats of mobility traces (*TrackReader*), so if a new format is needed only the appropriate

Table 1. Summary of configuration parameters.

Parameter	Description	Default
General configuration parameters		
-mcr	Mobile communication range	250 m
-v	Number of vehicles to simulate	100
-vld	Vehicle linear density (vehicles/Km). Overrides-v	
-s	Average speed of vehicles (Km/h)	50
-lat	Latency for a mobile agent's trip	1 s
-errRate	Error rate for communications	0%
-batch	Execute in batch mode	false
-map	Load a scenario map	Last used
-mbx	Number of *mailboxes* connected through a wired network (static nodes with storage capacity and the capability to provide wide-area network coverage)	0
-f	Number of fixed (non-moving) wireless devices	1
-mob	Mobility strategy for vehicles (1 = Random, 2 = Routes, 3 = Straight, 4 = Heuristic and Destination, 5 = GaussMarkov, 6 = ManhattanDynamic)	2
-nobldg	Do not simulate buildings on the map as communication obstacles	false
-rec	Record the experiment to a file for later replay and analysis	
-seed	Seed for random numbers (0 for random seed)	0
Configuration parameters to evaluate monitoring approaches		
-d	Initial distance (in meters) to the interest area	1000
-j	Mobile agent hop strategy (1 = RND, 2 = BEP, 3 = Approach, 4 = ANG, 5 = MAP, 6 = EP, 7 = EUC, 8 = Map-Traj, 9 = Optimal)	8
-m	Maximum number of simulation iterations during which to collect data	899
-mih	Minimum expected improvement to enable an agent's hop	0%
-dimArea	Dimensions of the interest area (height, width, longitude, and latitude: -mzh, -mzw, -mzw, -mzy)	
-prd	Probability that a device contains relevant data/appropriate sensors	50%
-rph	Max distance to hop (as a percentage of the communication range)	100%
-snru	Selection of nearest *Roadside Unit* (0 = none, 1 = once, 2 = continuous)	0

subclass must be implemented. Finally, one particularly important abstract class and interface are *Agent* and *IAgent*, respectively; they are intended to allow the programing of new different types of simulated mobile agents.

In this way, the simulator developed can be easily extended in a number of ways, such as: to incorporate new behaviors for the vehicles, e.g., new mobility strategies, to define new types of mobile agents with different behaviors (e.g., different agent mobility strategies), to add new simulation parameters, or to define different phases

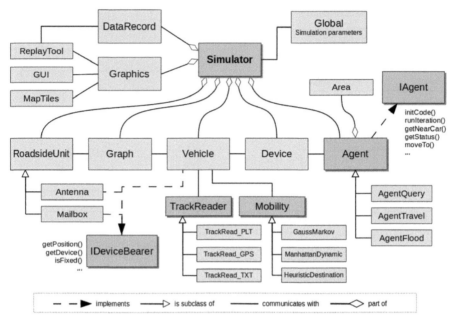

Figure 4. High-level overview of the architecture of the simulator.

in the data management approach for easy analysis of each of them with the replay tool; for example, phases such as "going to the area", "measuring environment data", and "coming back to the query originator" are defined in the monitoring task scenario, as shown in the upper-left part of Figure 1. We could also simulate data management approaches that do not use mobile agents, as we can simulate that there is a single static agent, i.e., a mobile agent with no mobility in each equipped vehicle that implements the required data management functions available in the equipped vehicles. Most extensions require defining appropriate subclasses and configuration parameters. Of course, for a comfortable use of the extensions in an interactive simulation mode, changes to the graphical user interface may also be required.

3.3 Examples of Simulation Scenarios

In this section, we describe some examples of scenarios that could be represented in the simulator in order to evaluate the potential interest of using different data management solutions based on the use of mobile agents with different behavior.

Example scenario 1: environmental monitoring in a city area

In this scenario, we monitor the environment using mobile agents. The idea is to exploit sensors available in conventional vehicles to monitor the environment, using mobile agents to find the relevant cars [26]. A *target area* (or *interest area*) is monitored, e.g., see Figure 5, which shows the target area as a rectangle near the center of the map, the node requesting the monitoring task as a point, and its communication range as

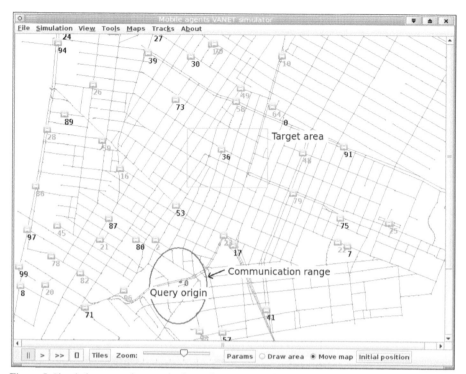

Figure 5. Simulating agents in a monitoring task: target area (blue rectangle) and query originator (center of the red circle).

a circle around the query point. One or more mobile agents can be sent to that area so that they can use sensors available in the vehicles to measure some environmental parameter, e.g., carbon monoxide, temperature, etc. in the area.

Thus, normal vehicles with the appropriate sensors are exploited in a dynamic way by using mobile agents, leading to a more flexible and economic solution than an alternative and more traditional approach where static sensors are manually deployed within the area of interest. However, the approach based on mobile agents faces some challenges, as the agents need to reach the target area by hopping from car to car (*transportation via communication*, using short-range wireless devices) and/or benefiting from the physical mobility of the vehicles (*transportation via locomotion*), keep themselves in vehicles within the area (to take the required measures), and finally return the results to the query originator. So, a key issue is to develop appropriate *hopping strategies* for the agents.

As we could imagine the development of different hopping strategies, we would like to be able to evaluate the performance of those strategies to know which one is the best, under what circumstances, and how long it takes for the entire monitoring process. The use of MAVSIM eases the process of designing and testing the behavior of the different strategies and the agents' themselves, thanks to the tight integration of all the simulated entities (moving vehicles, mobile agents, etc.), their interactions (the *communicates with* in the class diagram of Figure 4), and the facilities available for recording their status at every stage of the process (e.g., the device and vehicle where the agent is, if an agent hop ends successfully or fails, etc.). In this way, they can be analyzed later and/or reviewed with the replay tool, searching for any anomaly or pattern. If it were necessary to correct the behavior of the simulated agents, given the modularity of the simulator, only the agent code would need to be modified. The rest of the elements (vehicles, devices, etc.) would remain untouched.

Moreover, when a simulation ends different data are provided: the total number of iterations completed; if the simulated process ended successfully or not; the number of simulation iterations invested at every phase of the process; the number of times that every mobile agent hopped; their effective speed, etc. This data can be stored in a database or a spreadsheet, and if the simulation is repeated several times with variable initial conditions, the data collected can be analyzed to extract conclusions about the whole process performance. Since all the simulation parameters can be set from the command line and be executed unattended in batch mode, it is easy to automatize the execution of a high number of simulations varying only a few initial parameters, making the experimental data collection task very convenient.

Example scenario 2: research on climate variables

Another interesting scenario is monitoring moving regions, whose boundaries change over time. For example, we could envision the use of a VANET to help in the study of the possible correlation between storms and a higher presence of CO_2 in the atmosphere, as shown in a recent study [21].

This scenario is similar to the previous one we presented, but in this case the additional difficulty arises from the fact that the interest area is not a fixed set of coordinates in a city area, but a storm moving through a possibly larger area. Besides, the storm area could also change in size over time. In this case, the mobile agents would travel to the area by hopping from one car to another (by applying specific jumping protocols, as explained for the previous example scenario), but instead of determining if they are within the interest area by evaluating if their geographic location is within the boundaries of a fixed area, they could instead compute if a certain *area predicate* is satisfied in that location: in this case, a predicate that, based on environment data provided by sensors in the car, e.g., rain sensors automatically start the windshield wipers when it rains, can lead to an inference that the car is in a storm area. This strategy is suitable cars where the agents could take measures about the CO_2 and other target environment variables (by using the sensors available in those cars) could be those cars for which the area predicate that defines a storm has been satisfied recently. By using several mobile agents, we could also consider the problem of monitoring the current size and shape of the storm: mobile agents could try to keep close to the edges of the storm to keep track of its limits.

Another alternative would be to rely on some mechanism that keeps track of the current boundaries of the storm area, e.g., an external database or server, to update the mobile agents' knowledge about the interest area. Nevertheless, communicating the up-to-date area to the mobile agents would be a data management challenge if only short-range ad hoc communications (not wide-area communications) are available. It should be noted that it is not our purpose to show a full-fledged solution for this problem but just to illustrate another scenario which could be represented in the simulator. So, the simulator can be used to test a wide-area map, e.g., 10000 Km^2, simulating the movement of the storm using climate models or trajectories as simple or complex as needed; for this purpose, the *Area* class presented in Figure 4 can be extended with any behavior desired in order to enable moving interest areas.

4. Performance Evaluation of the Simulator

In this section, we present several tests that we have performed to evaluate the behavior of the simulator, as well as its throughput, when it executes simulations in scenarios with different numbers of entities and with varied complexity. Firstly, we evaluate the scalability of the simulator when we increase the number of vehicles. Secondly, we evaluate the scalability when the number of mobile agents in the tested application increases. Finally, we evaluate the simulation overhead when buildings are simulated.

The tests were executed using a computer with the following features: Linux Cent OS 5.11 operating system, with 16 GB of RAM and one six-core Intel Xeon X5660 processor running at 2.80 GHz. The simulator described in this chapter does not attempt to perform real-time simulations, as it uses an internal clock based on iterations of simulation, rather than a global real-time clock; by default, each iteration in the simulator represents 1 second of changes in the real-world. We do not include among the tests any experiment to evaluate the simulation of cognitive network because our current simulator can only be used to simulate the high-level effects of cognitive networks (by appropriately setting high-level parameters like those shown in Table 1).

4.1 Scalability with the Number of Vehicles

This test measures the speed (in iterations per second) achieved by the simulator with an increasing number of vehicles moving in an urban scenario, with the goal of finding out how well the simulator scales with the number of vehicles.

The selected scenario is a portion of the city of San Francisco (see Figure 6) extracted from OpenStreetMap. A varying number of vehicles traveling along its roads and streets are simulated by following a Random way-point mobility model [14], considering an average speed of 50 Km/h. The number of vehicles goes from 1000 to 10000 in intervals of 1000, and for each amount of vehicles the simulation is executed during 1000 iterations.

The results of the test can be seen in Figure 7, that shows the average of the results obtained with 50 repetitions of each test (in each test the initial random positions of the vehicles were different). In the figure, we can see how the speed of the simulator decreases as the number of vehicles increases. This was expected, as a higher number of vehicles implies a higher simulation overload. However, it can be noticed that the performance decrement is not linear, since the slope is steeper for values of less than 5000 vehicles and flatter for a higher number of vehicles. Thus, the results follow a logarithmic curve, which means that for more than 5000 vehicles the difference in terms of the simulation speed is smaller.

Figure 6. Scenario map of San Francisco.

The results obtained show that the overhead of the simulations is kept under control. In the worst case, for the evaluated scenario, we obtain a performance of around 10 iterations per second, which shows (if we assume that each iteration computes 1 second of changes, i.e., that it is equivalent to 1 second in the real work, as it is the default in our simulator) that the simulator could also be used for real-time simulations.

We have also measured the RAM needed to simulate a scenario with an increasing number of vehicles. During each simulation, the amount of RAM used by the Java process was measured, keeping track of the highest memory usage values during the execution of the simulation (lower values were discarded). As before, each experiment was repeated 10 times and an average delay for all the simulations was computed. The results of the test can be seen in Figure 8, which shows how the RAM needed increases with the number of vehicles, from approximately 1.5 GB for 1000 vehicles to almost exactly 8 GB for 10000 vehicles, following a linear trend.

4.2 Scalability with the Number of Mobile Agents

In this test, we measure the speed of the simulator (in iterations per second) depending on the number of simultaneous mobile agents that are simulated in a given scenario. The goal is thus to assess the scalability of the simulator in terms of the number of mobile agents in the application/system being tested.

As in Section 4.1, the scenario map is that of the city of San Francisco, and the number of vehicles is set to 1000. The number of mobile agents tested varies from 10 to 100 in increments of 10, and for each amount of agents the simulation is executed

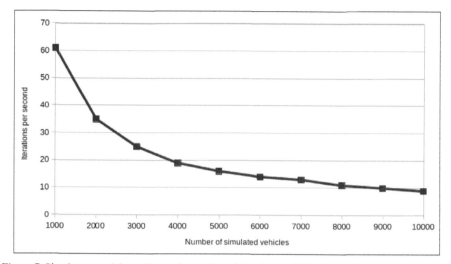

Figure 7. Simulator speed depending on the number of simulated vehicles.

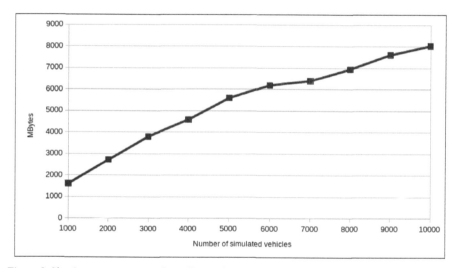

Figure 8. Simulator memory usage depending on the number of simulated vehicles.

during 1000 iterations. Each test is repeated 50 times with different initial random vehicle positions, and the average simulation speed is computed from all the obtained values.

The simulated agents behave in a way similar to that described in Section 3.3, that is they try to reach an interest area by hopping from one vehicle to another, in such a way that with every hop the agent moves closer to the area. This implies that the mobile agents are constantly evaluating the suitability of all the nearby surrounding vehicles, regarding the possibility to reach the interest area faster than with the vehicle where the agent is currently traveling. Moreover, when a mobile agent is within the interest area, it processes some data stored locally on the vehicles within the area. For the simulation to be more realistic, the presence of buildings that block communication signals are also simulated, for which the simulator must perform additional computations which slow down the simulation speed (as shown in Section 4.3).

The results of the test can be seen in Figure 9, where we can see how the speed of the simulator decreases as the number of simulated mobile agents increases. This decrement is not linear, since the slope is steeper for values of less than 40 agents, and flatter for amounts higher than that number. In any case, what is particularly important is that the overhead of mobile agents is not significant and that the system scales well with the number of mobile agents used.

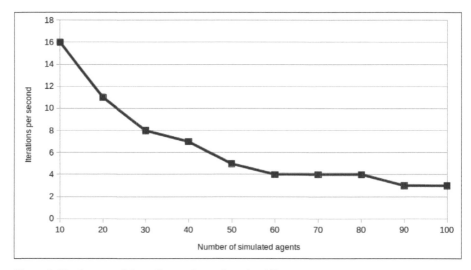

Figure 9. Simulator speed depending on the number of mobile agents.

We have also measured the RAM needed to simulate a scenario with an increasing number of mobile agents. During each simulation, the RAM used by the Java process was measured, keeping track of the highest memory usage values during the execution of the simulation (lower values were discarded). As before, each experiment was repeated 50 times and an average delay for all the simulations was computed. The results of the test can be seen in Figure 10, which shows how the RAM needed increases with the number of mobile agents, from approximately 3.3 GB for 10 agents, to about 3.4 GB for 100 agents. The total size may seem high, but it must be remembered that there are not only mobile agents data in the simulation but also the vehicles, graph map and others. Thus, taking only into account the mobile agent sizes from the test, we can calculate that each agent has an approximate size of 0.7 MB.

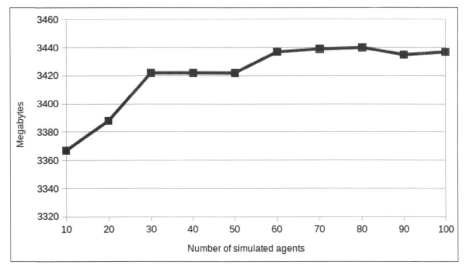

Figure 10. Simulator memory usage depending on the number of mobile agents.

4.3 Simulation Speed Depending on Whether Buildings are Simulated or Not

This test measures the speed (in iterations per second) achieved by the simulator depending on whether buildings are simulated or not, with the goal of finding out if the simulation of buildings implies a significant overhead or not. This study is motivated by the fact that the algorithm used to determine if a communication signal is blocked by buildings needs to compute a number of geometric equations [16] that can affect the amount of time that every iteration takes to complete.

For this test, we simulate a fixed number of vehicles (5000) and a mobile agent that tries to reach an interest area by hopping from one car to another and that processes some data stored locally on the vehicles within the area (as described in Section 3.3). If the mobile agent infers that a nearby car is a better candidate to travel towards the intended area than the current one that is physically carrying it, it will try to transfer itself to that other vehicle. However, the transmission of the agent could fail if, for example, the target car is behind a building that blocks the signal. Simulating the presence of a mobile agent or some type of communication between vehicles is necessary for this test, as in the absence of communications computing the presence or not of buildings is unnecessary (buildings have an impact only on the performance of communications, as they may block communication signals).

For this test, we execute 1000 iterations using three different maps, with different street layouts that can affect the complexity of computing the impact of buildings. The scenarios are fragments extracted from the maps of the following cities:

- London, which has an old-city street layout with short, narrow and curved streets.
- New York, with long and straight streets in a grid-type layout.
- San Francisco, which is a mixed-style city that has both types of streets.

A fragment of the maps of London and New York, used in the experiments, are shown in Figure 11 and Figure 12, respectively. The map fragment for San Francisco can be seen in Figure 6.

For each map, we repeat the simulation with buildings and without buildings and with 50 different random initial vehicle positions. The average of the simulation speeds is computed and shown in Figure 13.

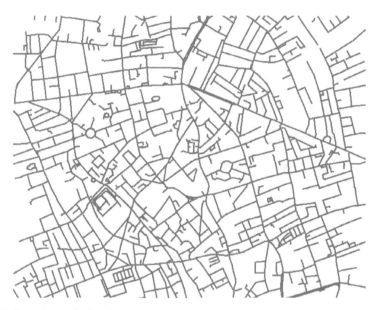

Figure 11. Scenario map for London.

Figure 12. Scenario maps for New York.

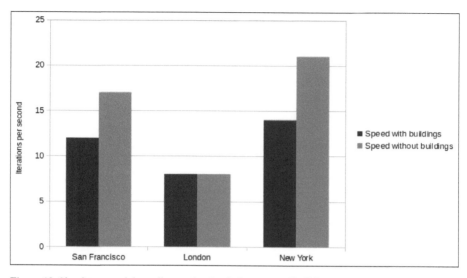

Figure 13. Simulator speed depending on the simulation or not of buildings.

We can see that in general the simulation speed is lower when the presence of buildings is simulated, as it was expected, due to the overhead of computing the equations necessary to evaluate the effect of buildings. A notable exception is the case of London, where both values are very similar; this is due to the presence of short and curved streets, which allows the algorithm for buildings used by the simulator to determine at a very early stage that the communication is not possible, without hardly performing additional calculations.

We can also analyze the impact of the city topology on the simulation speed. The performance of simulations in the city with straighter layout (New York) is higher than in the city with a more complicated arrangement (London).

We also note that, independently of whether buildings are simulated or not, if we compare the results obtained in the test presented in Section 4.1 for 5000 vehicles and the case of San Francisco with the equivalent results shown in this section, the simulation speed is a bit lower for the experiment described in this section. The reason is that when only moving vehicles are being simulated (as in Section 4.1) we avoid the additional overhead due to the execution of the mobile agent, which constantly evaluates the convenience of the vehicle which is currently carrying it, i.e., keeps itself on alert looking for a better "taxi" to hop to.

5. Conclusions and Future Work

Vehicular networks bring important challenges for the data management community. For practical reasons, the proposed communication and data management strategies are usually evaluated using simulators, which provides flexibility in defining the required test conditions and enables quick results considering large-scale simulated scenarios with an affordable cost. Although the use of mobile agents in the context of VANETs could be interesting, existing simulators cannot be easily adapted to evaluate approaches based on the use of mobile agents. For that reason, we have developed our own simulator.

We think that the simulator developed is well suited for testing applications, protocols and systems that require the use of mobile agents in vehicular networks, taking into account different elements such as the network communications, the moving vehicles, the presence of buildings that might block the signal, and other parameters of interest for testing different scenarios. The simulator is written in a modular and extensible way, so it can be enhanced with new features with little effort.

We are currently using the simulator in our research, and we are constantly adding new features when we need them to evaluate different data management strategies. Thanks to the use of this simulator, we have obtained promising results regarding the potential interest of using mobile agents in vehicular networks [26, 27, 25].

There are, however, a number of features and enhancements that have not been implemented so far. Among these, we are especially interested in the development of some extensions to simulate *events*, e.g., accidents, traffic jams, etc., and scarce resources (particularly, *parking spaces*), as they will enable testing data management strategies for the exchange of interesting events for drivers [4] and applications to help drivers in the search of appropriate parking spaces [8]. Moreover, we also plan to study the possibility of mixing the simulator developed with functionalities provided by existing general-purpose traffic simulators (in particular, *SUMO* with *TraCI*) and lower-level network simulators (such as NS-2). For example, our current simulator can be used to simulate the high-level effects of cognitive networks, e.g., better usage of the communication spectrum, which can lead to less transmission failures, but by combining it with specialized network simulators it would be possible to simulate also the low-level aspects in a reliable way.

Acknowledgments

This work has been supported by the CICYT project TIN2013-46238-C4-4-R and DGA-FSE. We thank Eduardo López for his previous contribution to an old prototype of the simulator.

References

[1] A.K. Al-Ali, Yifan Sun, M. Di Felice, J. Paavola and K.R. Chowdhury. Accessing spectrum databases using interference alignment in vehicular cognitive radio networks. IEEE Transactions on Vehicular Technology, 64(1): 263–272, January 2015.

[2] A. Bonifati, P.K. Chrysanthis, A.M. Ouksel and K. Sattler. Distributed databases and peer-to-peer databases: past and present. SIGMOD Record, 37: 5–11, March 2008.

[3] M.J. Booysen, S. Zeadally and G.J. van Rooyen. Performance comparison of media access control protocols for vehicular ad hoc networks. IET networks, 1(1): 10–19, March 2012.

[4] N. Cenerario, T. Delot and S. Ilarri. A content-based dissemination protocol for VANETs: Exploiting the encounter probability. IEEE Transactions on Intelligent Transportation Systems, 12(3): 771–782, September 2011.

[5] T. Delot and S. Ilarri. Data gathering in vehicular networks: The VESPA experience (invited paper). In Fifth IEEE Workshop On User MObility and VEhicular Networks (ON-MOVE 2011), at LCN 2011, pp. 801–808. IEEE, 2011.

[6] T. Delot and S. Ilarri. Introduction to the special issue on data management in vehicular networks. Transportation Research Part C: Emerging Technologies, 23: 1–2, August 2012.

[7] T. Delot and S. Ilarri. The VESPA Project: Driving advances in data management for vehicular networks. ERCIM News, (94): 17–18, 2013. Special Theme on Intelligent Vehicles as an Integral Part of Intelligent Transport Systems.

[8] T. Delot, S. Ilarri, S. Lecomte and N. Cenerario. Sharing with caution: Managing parking spaces in vehicular networks. Mobile Information Systems, 9(1): 69–98, February 2013.

[9] M. Di Felice, K.R. Chowdhury, W. Kim, A. Kassler and L. Bononi. End-to-end protocols for cognitive radio ad hoc networks: An evaluation study. Performance Evaluation, 68(9): 859–875, September 2011.

[10] D. Hamza, K. Park, M. Alouini and S. Aissa. Throughput maximization for cognitive radio networks using active cooperation and superposition coding. IEEE Transactions on Wireless Communications, 14(6): 3322–3336, June 2015.

[11] J. Harri, F. Filali and C. Bonnet. Mobility models for vehicular ad hoc networks: A survey and taxonomy. IEEE Communications Surveys & Tutorials, 11(4): 19–41, October 2009.

[12] IEEE. IEEE standard for policy language requirements and system architectures for dynamic spectrum access systems. IEEE Std 1900.5-2011, pp. 1–51, January 2012.

[13] S. Ilarri, R. Trillo and E. Mena. SPRINGS: A scalable platform for highly mobile agents in distributed computing environments. In Fourth International Workshop on Mobile Distributed Computing (MDC'06), at WoWMoM'06, pp. 633–637. IEEE, 2006.

[14] V.D. Khairnar and S.N. Pradhan. Mobility models for vehicular ad-hoc network simulation. In IEEE Symposium on Computers Informatics (ISCI 2011), pp. 460–465, 2011.

[15] S. Li, Z. Zheng, E. Ekici and N. Shroff. Maximizing system throughput by cooperative sensing in cognitive radio networks. IEEE/ACM Transactions on Networking, 22(4): 1245–1256, August 2014.

[16] F.J. Martinez, M. Fogue, C.K. Toh, J.C. Cano, C.T. Calafate and P. Manzoni. Computer simulations of VANETs using realistic city topologies. Wireless Personal Communications, 69(2): 639–663, March 2013.

[17] F.J. Martinez, C.K. Toh, J.C. Cano, C.T. Calafate and P. Manzoni. A survey and comparative study of simulators for vehicular ad hoc networks (VANETs). Wireless Communications & Mobile Computing, 11(7): 813–828, July 2011.

[18] D. Milojicic, F. Douglis and R. Wheeler. Mobility: Processes, Computers, and Agents. ACM, 1999.

[19] J. Mitola. Cognitive Radio—An Integrated Agent Architecture for Software Defined Radio. PhD thesis, Royal Institute of Technology (KTH), 2000.

[20] S. Olariu and M.C. Weigle, editors. Vehicular Networks: From Theory to Practice. Chapman & Hall/ CRC, 2009.

[21] M. Reichstein, M. Bahn, P. Ciais, D. Frank, M.D. Mahecha, S.I. Seneviratne, J. Zscheischler, C. Beer, N. Buchmann, D.C. Frank et al. Climate extremes and the carbon cycle. Nature, 500(7462): 287–295, August 2013.

[22] R. Sengupta and S. Bhattacharjee. Comparative analysis of GA and SA for utility maximization of licensed and unlicensed users in a cognitive radio network. In International Conference on Control, Instrumentation, Energy and Communication (CIEC 2014), pp. 498–502, 2014.

[23] J. Treurniet. A taxonomy and survey of microscopic mobility models from the mobile networking domain. ACM Computing Surveys, 47(1): 14:1–14:32, July 2014.

[24] R. Trillo, S. Ilarri and E. Mena. Comparison and performance evaluation of mobile agent platforms. In Third International Conference on Autonomic and Autonomous Systems (ICAS'07). IEEE, 2007.

[25] O. Urra and S. Ilarri. Using mobile agents in vehicular networks for data processing. In 14th International Conference on Mobile Data Management (MDM 2013), 2: 11–14. IEEE, 2013.

[26] O. Urra, S. Ilarri, T. Delot and E. Mena. Using hitchhiker mobile agents for environment monitoring. In Seventh International Conference on Practical Applications of Agents and Multi-Agent Systems (PAAMS'09), volume 55 of Advances in Intelligent and Soft Computing, pp. 557–566. Springer, 2009.

[27] O. Urra, S. Ilarri, T. Delot and E. Mena. Mobile agents in vehicular networks: Taking a first ride. In Eight International Conference on Practical Applications of Agents and Multi-Agent Systems (PAAMS 2010), volume 70 of Advances in Intelligent and Soft Computing, pp. 118–124. Springer, 2010.

[28] O. Urra, S. Ilarri and E. López. Context-aware recommendations in mobile environments. In XIX Jornadas de Ingeniería del Software y Bases de Datos (JISBD 2014), Cádiz (Spain), pp. 71–84. ISBN-13: 978-84-697-1152-1, ISBN-10: 84-697-1152-0, 2014.

[29] J. Zhao and G. Cao. VADD: Vehicle-assisted data delivery in vehicular ad hoc networks. IEEE Transactions on Vehicular Technology, 57(3): 1910–1922, May 2008.

[30] M.T. Zia, F.F. Qureshi and S.S. Shah. Energy efficient cognitive radio MAC protocols for ad hoc network: A survey. In 15th International Conference on Computer Modelling and Simulation (UKSim 2013), pp. 140–143, 2013.

Index

An environmentally friendly book printed and bound in England by www.printondemand-worldwide.com